CELL ATP

William A. Bridger

J. Frank Henderson

The University of Alberta
Edmonton, Canada

A Wiley-Interscience Publication
JOHN WILEY & SONS
New York • Chichester • Brisbane • Toronto • Singapore

Library of Congress Cataloging in Publication Data:

Bridger, William A.
Cell ATP.

(Transport in life science series, ISSN 0271-6208; v. 5)
"A Wiley-Interscience publication."
Includes index.
1. Adenosine triphosphate. 2. Adenosine triphosphate —Metabolism. 3. Cell metabolism. I. Henderson, J. Frank (Joseph Frank), 1933– II. Series.
[DNLM: 1. Adenosine triphosphate—Physiology. 2. Cells —Physiology. 3. Energy metabolism. W1 TR 235V v.5 / QU 58 B851c]

QP625.A3B74 1983 591.87′328 82-24797
ISBN 0-471-08507-3

Printed in the United States of America

10 9 8 7 6 5 4 3 2 1

SERIES PREFACE

Membrane transport is rapidly becoming one of the best-worked fields of modern cell biology. The Transport in the Life Sciences series deals with this broad subject in monograph form. Each monograph seeks to trace the origin and development of ideas in the subject in such a way as to show its true relation to membrane function. It also seeks to present an up-to-date and readable outline of the main problems in the subject and to guide thought on to new lines of investigation.

The task of writing a monograph is not a light one. My special gratitude is to the various authors for expounding their subjects with scholarly care and force. For the preparation of the indexes I thank Dr. Barbara Littlewood.

E. EDWARD BITTAR

Madison, Wisconsin
June 1980

PREFACE

Most processes of membrane transport depend, directly or indirectly, on ATP as the principal mediator of cellular energy metabolism. It is therefore appropriate for the Transport in the Life Science series to include a monograph in which the basic features of the structure, function, and metabolism of ATP are reviewed. However, because the reactions of ATP metabolism and function are not limited only to the transport processes, the contents of this volume are not specifically related at every stage to the relation of ATP to transport, and they are intended to be of interest as well to a wide variety of investigators.

Our consideration of ATP falls naturally into two parts, that relating to the phosphoryl moieties and that relating to the adenine and ribose portions of the ATP molecule. In preparing this volume, we have indeed been impressed by how distinct these two aspects of ATP metabolism and function really are, and although we have read and commented on each other's material, the two halves of the book have been written relatively independently. W.A.B. has prepared the first five chapters, which deal primarily with the chemistry and metabolism of the phosphoryl residues of the ATP molecule. This includes detailed consideration of the mechanisms of action of enzymes that catalyze ATP utilization by processes including membrane transport, the mechanisms of rephosphorylation of ADP to produce ATP by cellular systems such as oxidative phosphorylation, and review of the role that ATP may play as a metabolic regulator. J.F.H. has written the last five chapters, which deal with the adenosine moiety, and has organized the material around the following questions: What factors influence the concentration of ATP in cells? What factors, whether physiological, pharmacological, or pathological, can serve to increase ATP concentrations, and which can serve to decrease them? The relatively comprehensive nature of our discussion of ATP should thus be evident, and we hope that interest to students of transport processes will be apparent.

William A. Bridger

J. Frank Henderson

Edmonton, Canada
April 1983

CONTENTS

1 Introduction **1**

Historical Introduction, 1
ATP Concentrations in Cells and Tissues, 4

PART I

2 Chemistry of ATP **9**

Site of Bond Cleavage, 9
Standard Free Energy of ATP Hydrolysis, 10
Coordination of Metal Ions, 15

3 Mechanisms of ATP Production **20**

Substrate-Level Phosphorylations, 20
Electron Transport-Linked ATP Synthesis, 29

4 Mechanisms of ATP Utilization **41**

Sodium–Potassium ATPase, 41
ATP Hydrolysis by Myosin and Actomyosin, 46
Hexokinase, 49

5 ATP and Metabolic Regulation **57**

Maintenance of ATP Concentration, 58
Protein Kinases and Metabolic Control, 60
Regulation of Phosphofructokinase Activity, 60

PART II

6 ATP Synthesis **69**

Basic Pathways of ATP Synthesis, 69
Regulation of ATP Synthesis as a Whole, 73

7 ATP Catabolism **94**

Pathways of ATP Catabolism, 94
Regulation of ATP Catabolism as a Whole, 97

8 Metabolism of the Adenosine Moiety of ATP in Intact Cells and Animals **114**

Basic Functions of ATP and Its Utilization at the Cellular Level, 115
Factors Affecting ATP Metabolism in Tissues and Animals, 118
Functions of ATP and Its Utilization in Tissues and Animals, 122
Cyclic Pathways of ATP Utilization and Function, 124

9 Pathological Influence on ATP Metabolism **135**

Elevated ATP Concentration, 135
Lowered ATP Concentration, 140

10 Effects of Drugs **145**

Elevation of ATP Concentrations, 145
Lowering of ATP Concentrations, 152

Index **163**

CELL ATP

1

Introduction

HISTORICAL INTRODUCTION

Morowitz[1] states:

> The act of running is linked to that subtle balance of energy by which we ingest aliment, maintain our thermodynamically unstable systems, and metaphorically speaking, engage in the chase, our ultimate animal function. The final step in each and every one of these biological processes is the transformation of energy from ATP to its ultimately usable form. In the muscle this conversion changes the chemical energy of ATP into mechanical work directed along the actomyosin fibers. It is a process not yet completely understood, yet one that makes possible every motion that we undertake. The brain wills and the muscle contracts—it is the physiological essence of the mind–body problem. Doubtlessly one of these days a poet will appear to honor the molecule of ATP. Until then we can only celebrate this vital part of us in the somewhat cold, formal language of biochemistry. As a well-known sports announcer would have put it, ATP is the most underrated molecule in the league today.

Adenosine triphosphate (ATP) has an utterly central role in energy metabolism and, therefore, in biology. This concept, now fundamental to any biochemist's credo, seems to be surprisingly recent. Adenosine triphosphate was not known until 1929, when Fiske and Subbarow[2] working in the United States and Lohmann[3] in Germany independently detected the substance among the phosphorylated components of muscle extracts. The importance of ATP was not immediately recognized, partly because of its low concentration relative to that of creatine phosphate, and also because the scientific community at the time was being deluged with discoveries of labile phosphate compounds, including creatine phosphate, inorganic pyrophosphate, and myoadenylic acid [adenosine 5′-monophosphate (5′-AMP)]. The chemical structure of ATP (Fig. 1.1) was proposed shortly thereafter by Lohmann[4] and confirmed in 1948 by Lord Todd at Cambridge, who achieved total chemical synthesis.[5]

The 1930s witnessed a series of separate discoveries that together formed the framework for recognition of the fundamental role of ATP and its relationship to energy production through the Embden–Meyerhof pathway. Working together with Meyerhof, Lohmann discovered the "energy-rich" nature of both phosphate anhydride linkages of ATP and made the first estimates of the energy released by their cleavage.[6,7] This was accompanied by Lohmann's discovery of creatine kinase activity in muscle, establishing the possibility of metabolic equivalence of the phosphoryl groups of ATP and that of creatine phosphate.[8] This, together with the finding that the newly discovered myosin catalyzes the hydrolysis of ATP,[9] suggested that ATP hydrolysis could serve as the primary source of energy for muscular contraction. During this period, separate investigations of the glycolytic pathway in cell-free systems led to the important conclusion that ATP synthesis was a by-product of glucose utilization. Karl Lohmann again played one of the leading roles in the development of this concept; together with Mey-

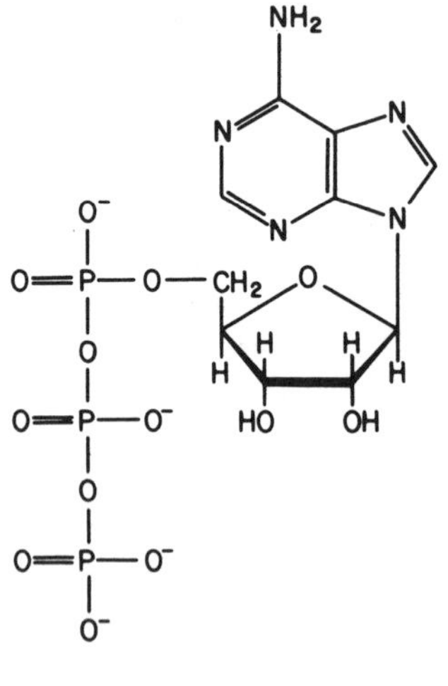

Figure 1.1. Chemical structure of ATP. The tetraanionic form shown is the predominant species at physiological pH (see Chapter 2).

erhof, Lohmann measured the stoichiometric coupling of ATP production to lactate accumulation in extracts of muscle carrying out glycolysis.[10] Prior to this time, it had been accepted that muscular work was directly coupled to lactate production by anaerobic glycolysis. One inkling that this prevalent concept would require modification came from the important observation by Lundsgaard[11] that muscle tissues can continue to contract even after they have been poisoned with sufficient iodoacetate to block glycolysis completely. Lundsgaard also showed that the contraction of iodoacetate-treated muscle was accompanied by the dephosphorylation of ATP and creatine phosphate and that their rephosphorylation could take place during aerobic oxidation of lactate. The energy-releasing dephosphorylations could thus be viewed in terms of providing energy for muscular contraction, whereas the replenishment of these phosphorylated compounds could take place separately from contraction per se by the energy released by glycolysis or aerobic oxidation of lactate. Lohmann and Meyerhof's work, together with the major contributions of Parnas et al.,[12] established that conversion of hexose phosphates to pyruvate required the presence of a phosphate acceptor, namely, AMP or adenosine diphosphate (ADP). Korzybski and Parnas[13] thus formulated a "phosphate cycle" whereby the investment of one ATP for the phosphorylation of hexose is counterbalanced by the regeneration of ATP during subsequent steps of glycolysis, but the concept of ATP as the principal currency of cellular energy had not yet fully crystallized.

In 1931, Meyerhof and Lohmann[6] speculated on the possible utilization of ATP bond energy, captured by glycolysis, in biosynthetic processes. This concept and the phosphate cycle described by Parnas received little attention until the timely appearance of a classic review article by Fritz Lipmann in the first volume of *Advances in Enzymology*.[14] Here Lipmann introduced the "squiggle" (~) to the biochemist's vocabulary. His formulation of the ATP metabolic wheel, shown in Figure 1.2 represents a watershed in our understanding of the interplay of catabolic and anabolic processes and of the central role of ATP as the basic currency of metabolism. Lipmann explicitly introduced the concept of the preservation of bond energy. It had long been recognized that there were at least two classes of phosphorylated compound in biological extracts that differ significantly in their thermodynamic stability as reflected by their $\Delta G^{\circ\prime}$ values for hydrolysis. For simple phosphate esters such as glucose-6-phosphate, this value is approximately -3 kcal/mole, whereas for compounds such as ATP, creatine phosphate, arginine phosphate, phosphoenolpyruvate, and acetyl phosphate, the $\Delta G^{\circ\prime}$ value for hydrolysis was found to be near -10 kcal/mole. Lipmann's squiggle ~ was introduced to designate this so-called high-energy bond, char-

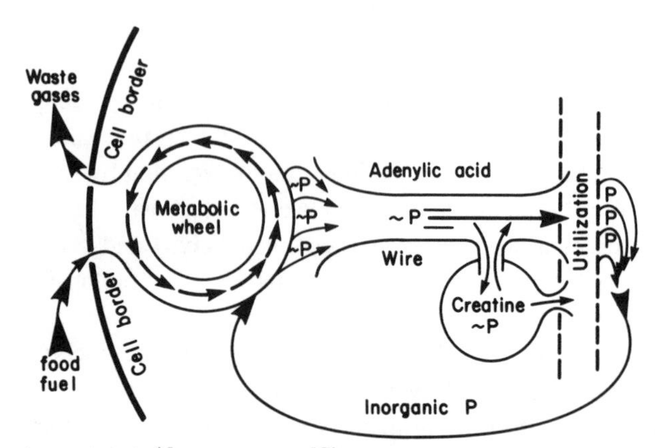

Figure 1.2. The cycle of ATP production and utilization as originally formulated by Fritz Lipmann.[14] This was the first clear enunciation of the principle that the catabolism of fuel molecules such as glucose to their waste products is responsible for the capture of conserved energy by means of the conversion of inorganic phosphate to phosphorylated adenylic acid (ATP). It is also particularly significant that Lipmann visualized ATP as the primary form of chemical bond energy for direct utilization, with creatine phosphate serving a secondary role as a storage form for phosphoanhydride bond energy.

acterized by a chemical potential higher than that found in phosphate esters. The metabolic significance of reversible formation of ~P compounds soon became evident—the dependence of chemical or physical work on metabolism becomes understandable because the latter generates ATP.

For a more extended review of the discovery of ATP and the development of the concept of ATP as the primary form of usable chemical bond energy in the cell, the reader is referred to Marcel Florkin's *A History of Biochemistry,*[15] particularly Chapters 23 and 25, and to Lipmann's autobiographical narrative *Wanderings of a Biochemist.*[16]

ATP CONCENTRATIONS IN CELLS AND TISSUES

Adenosine triphosphate is present in most cells in substantial amounts, and its concentrations are higher than those of other purine or pyrimidine nucleotides. However, ATP concentrations do or may vary appreciably among cell types, and for a given type, among species; in addition, they may change depending on physiological conditions, age, growth rate, and other factors. The literature on this subject is enormous,[17] and a thorough survey of this subject is not attempted here. However, a small number of examples may be given.

In cultured mammalian cells, ATP concentrations have been reported to vary from 2.4 to 11 nmol/10^6 cells (Snyder et al.[18] and J. Hordern and J. F. Henderson, unpublished results), with some of this variation due to differences in cell volume. Adenosine triphosphate levels per cell increase during progression through the cell cycle, but cell size also increases, and there is little change per unit volume.[18,19] Some other typical values include 4.7–7.8 μmol ATP per gram wet weight of skeletal muscle[20] and 4.5 μmol ATP per gram for heart.[21] Human erythrocytes contain 1–2 μmol of ATP per millimeter of packed cells,[22] but there is considerable species variation in the reported values, from 0.2 in horses[23] to 15 in snakes.[24] It has recently been reported that erythrocytes of the echidna contain no ATP at all![25]

Within cells, ATP can at least potentially be compartmented within mitochondria, nuclei, and other organelles, as well as bound to other cellular constituents. The quantitative analysis of such compartmentation is very difficult, and published estimates vary widely. In one recent study, concentrations of ATP in rat liver cytosol and mitochondria were reported to be 6.2 ± 0.63 and 7.5 ± 0.73 μmol/ml water, respectively.[26] It seems unlikely that there is any appreciable difference between nucleus and cytoplasm with respect to ATP concentrations. The whole subject of compartmentation requires much further study.

Under normal conditions, ATP is the quantitatively predominant adenine nucleotide, and the relationships among the concentrations of ATP, ADP, adenylate, and deoxy-ATP are of interest. In cultured animal cells, ATP:ADP ratios typically are in the range 8–12, and ATP:adenylate ratios are close to 50.[18,19,27] However, in animal cells *in vivo,* the ATP:ADP ratio may be lower (perhaps as low as 2–3 following severe energy demands), but whether these estimates reflect the true intracellular situation or are due to some breakdown of ATP during extraction is not certain. Finally, ATP:deoxy-ATP ratios range from 150 to 450 in different cultured cells (Hordern and Henderson, Ref. 19 and unpublished results). The concentration of ATP usually is higher than those of guanosine 5′-triphosphate (GTP), uridine triphosphate (UTP), and cytidine 5′-triphosphate (CTP), although the relative concentrations of all four nucleoside triphosphates vary considerably among cell types and tissues. For example, adenine nucleotides comprise approximately 90 percent of total nucleotides in skeletal muscle but as little as 50–60 percent in kidney. Such ratios can also vary with age and physiological conditions.[17]

REFERENCES

1. H. J. Morowitz, in *Bay to Breakers, The Wine of Life and Other Essays on Societies, Energy and Living Things,* St. Martin's Press, New York (1979), p. 20.
2. C. H. Fiske and Y. Subbarow, *Science* **70,** 381 (1929).
3. K. Lohmann, *Naturwissenschaften* **17,** 624 (1929).
4. K. Lohmann, *Biochem. Z.* **254,** 381 (1932).
5. B. Lythgoe and A. R. Todd, *Nature* **155,** 695 (1945).
6. O. Myerhof and K. Lohmann, *Naturwissenschaften* **19,** 575 (1931).
7. O. Myerhof and K. Lohmann, *Biochem. Z.* **253,** 431 (1932).
8. K. Lohmann, *Biochem. Z.* **271,** 264 (1932).
9. W. A. Engelhardt and M. N. Ljubimova, *Nature* **144,** 669 (1939).
10. K. Lohmann, *Biochem. Z.* **271,** 109, 120 (1934).
11. E. Lundsgaard, *Biochem. Z.* **217,** 162 (1936).
12. J. K. Parnas, P. Ostern, and T. Mann, *Biochem Z.* **272,** 64 (1934).
13. T. Korzybski and J. K. Parnas, *Bull. Soc. Chim. Biol.* **21,** 713 (1939).
14. F. Lipmann, *Adv. Enzymol.* **1,** 99 (1941).
15. M. Florkin, *A History of Biochemistry,* pt. III, vol. 31, *Comprehensive Biochemistry,* M. Florkin and E. H. Stotz, Eds., Elsevier, Amsterdam (1975).
16. F. Lipmann, *Wanderings of a Biochemist,* Wiley, New York (1971).
17. P. Mandel, *Prog. Nucl. Acid Res. Mol. Biol.* **3,** 299 (1964).
18. F. F. Snyder, J. F. Henderson, S. C. Kim, A. R. P. Paterson, and L. W. Brox, *Cancer Res.* **33,** 2425 (1973).
19. J. Hordern and J. F. Henderson, *Can. J. Biochem.* (in press).
20. R. A. Meyer, G. A. Dudley, and R. L. Terjung, *J. Appl. Physiol.* **49,** 1037 (1980).
21. H.-G. Zimmer, G. Steinkopff, H. Ibel, and H. Koschine, *J. Mol. Cell Cardiol.* **12,** 421 (1980).
22. D. Rubinstein and E. Warrendorf, *Can. J. Biochem.* **53,** 671 (1975).
23. T. J. McManus, in *The Human Red Cell in vitro,* T. J. Greenwalt and G. A. Jamieson, Eds., Grune and Stratton, New York (1974) p. 49.
24. S. Rapoport and G. M. Guest, *J. Biol. Chem.* **138,** 269 (1941).
25. H. D. Kim, R. B. Zeidler, J. D. Sallis, S. C. Nichol, and R. E. Isaacks, *Science* **213,** 1517 (1981).
26. W.-D. Schwenke, S. Soboll, H. J. Seitz, and H. Sies, *Biochem. J.* **200,** 405 (1981).
27. J. F. Henderson, J. K. Lowe, and J. Barankiewicz, in *Purine and Pyrimidine Metabolism,* Elsevier, Amsterdam (1977), p. 3.

PART I

2

Chemistry of ATP

The utilization of ATP by metabolic processes (biochemical syntheses, membrane transport, muscle contraction, etc.) involves the cleavage of the polyphosphate region of the ATP molecule. This chapter contains a survey of those aspects of the chemistry of ATP that are particularly relevant to these enzyme-catalyzed processes: the sites of bond cleavage; estimates of free energy changes associated with such cleavage; the dissociation constants for hydrogen and metal ions; and the sites of metal ion coordination. As is seen later, much of the most recent and most informative studies involve the application of ^{31}P-nuclear magnetic resonance (^{31}P-NMR) to the structure of ATP and its analogues; therefore, a summary of some important NMR parameters is also included.

SITE OF BOND CLEAVAGE

With the availability of ^{18}O-labeled water in the 1950s, techniques were developed, particularly by Mildred Cohn and Paul Boyer, for the analysis of the site of cleavage of ATP by various enzyme systems.[1] The results of numerous investigations, principally in these two laboratories, may be summarized in Figure 2.1.

Without known exception, the pattern of cleavage of the β–γ linkage involves

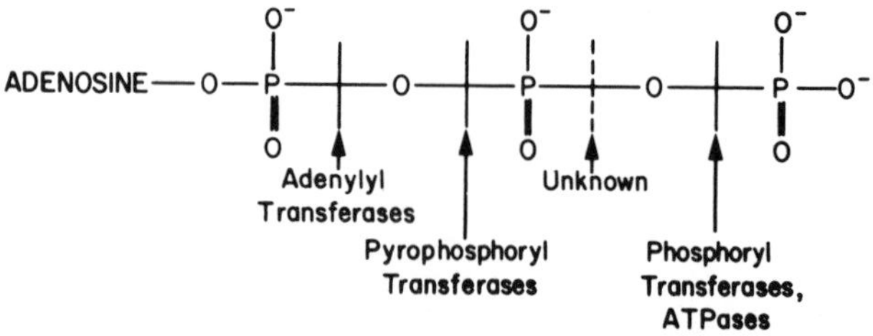

Figure 2.1. Known sites of ATP bond cleavage by various enzyme systems. (Adapted from Reference 1.)

breakage of the O–P_γ bond; the phosphoryl group thus liberated may be transferred directly to an acceptor molecule (water, glucose, creatine, etc.) or may be transiently attached to an amino acid side chain (serine, histidine, etc.) of a phosphoryl-transferring enzyme. Almost all enzymes that catalyze cleavage of the α–β linkage act as adenylyl transferases [acetyl–coenzyme (CoA) synthase, amino acyl-tRNA synthases, etc.], and accordingly the P_α–O bond is the bond that is broken. A small group of enzymes (phosphoenolpyruvate synthase and phosphate–pyruvate dikinase) act with the intermediate transfer of a pyrophosphoryl group to a histidyl group of the respective enzyme, and accordingly their catalysis involves the cleavage of the O–P_β bond.

STANDARD FREE ENERGY OF ATP HYDROLYSIS

The change in free energy that is associated with the cleavage or formation of either the α–β or β–γ bond of ATP is the common denominator that links exergonic processes such as glycolysis and oxidative or photosynthetic phosphorylation to energy-requiring processes such as biosynthesis of cellular constituents, membrane-active transport, and muscle contraction. The reaction for the hydrolysis of ATP to ADP at neutral pH is

$$ATP^{4-} + H_2O \rightleftharpoons ADP^{3-} + HPO_4^{2-} + H^+ \qquad (1)$$

The equilibrium for this reaction lies far to the right; a fundamental design concept of metabolism is to use the energy of this reaction by coupling it to drive unfavorable ones, such as endergonic biosynthetic reactions. Knowledge of the free energy change for ATP hydrolysis requires an accurate knowledge of the observed free energy change under the same conditions. The most direct approach to the experimental evaluation of this important quantity under a given set of conditions would be to measure the observed equilibrium constant for equation (1). However, the position of this equilibrium is so far to the right that an accurate measurement by this direct method is not technically feasible. Thus the equilibrium constant has been estimated from two or more reactions, the sum of which is equation (1). It must be borne in mind that experimental evaluation of this important quantity is further complicated by the very significant effects of Mg^{2+} and pH on the free energy change.

Many detailed attempts to estimate $\Delta G^{\circ\prime}$ have made use of the glutamine synthase and glutaminase reactions. These include the studies of Benzinger et al.,[2] who determined the equilibrium constant of the glutaminase reaction and combined this with the constant for glutamine synthase as estimated by Levintow and Meister.[3] Alberty[4] and Phillips et al.[5] have handled the same data differently (particularly with regard to the Mg^{2+} and pH effects) and have produced a different set of values. Furthermore, Rosing and Slater[6] have redetermined the equilibrium constant for glutamine synthase, leading to yet another discrepant estimate for $\Delta G^{\circ\prime}$. Subsequently, in view of these discrepancies, Guynn and Veech[7] have determined the equilibrium constant for equation (1) from the combined equilibrium of the acetate kinase and phosphate acetyltransferase reactions, together with knowledge of the $\Delta G^{\circ\prime}$ for acetyl–CoA hydrolysis.[8] Their estimates, together with those obtained with the glutamine synthase–glutaminase system, are summarized in Table 2.1. An evaluation of the experimental complications of the various methods for estimation of $\Delta G^{\circ\prime}$ suggests that the most appropriate value to adopt for physiological conditions is that given by Guynn and Veech[7] (pH 7.0, free Mg^{2+} = 1 m*M*):

$$\Delta G^{\circ\prime} = -7.6 \text{ kcal}(-31.8 \text{ kJ})/\text{mole}$$

Table 2.1. ΔG° Estimates for ATP Hydrolysis (pH 7.0, 37°C)

	Ref. 7	Ref. 2	Ref. 6	Ref. 5
$ATP^{4-} + H_2O \rightleftharpoons ADP^{3-} + HPO_4^{2-} + H^+$	8.41	8.6	7.9	9.9
$MgATP^{2-} + H_2O \rightleftharpoons MgADP^{-} + HPO_4^{2-} + H^+$	6.95	7.0	6.3	
$HATP^{3-} + H_2O \rightleftharpoons HADP^{2-} + H_2PO_4^{-}$	7.67		7.37	9.34
$\Sigma ATP + H_2O \rightleftharpoons \Sigma ADP + \Sigma P_i$[a]	8.53		7.93	10.06
$\Sigma ATP + H_2O \rightleftharpoons \Sigma ADP + \Sigma P_i$[b]	7.60		6.79	8.74

[a]Free $[Mg^{2+}] = 0$.
[b]Free $[Mg^{2+}] = 1$ m*M*.

As would be expected from the acid dissociation constants of the components of equation (1) and their dissociation constants for Mg^{2+}, the $\Delta G^{\circ\prime}$ is a strong function of pH and $[Mg^{2+}]$. The dramatic effect of the concentration of divalent cation is clearly illustrated in Figure 2.2,[7] and the complex nature of the variation of $\Delta G^{\circ\prime}$ with both of these parameters is illustrated by the contours calculated by Alberty[4] (Figure 2.3).

The free energies for hydrolysis of many naturally occurring phosphate-containing compounds are listed in Table 2.2.

Several important generalities may be made from these values:

1. Simple esters of phosphoric acid with primary and secondary alcohols (glucose-6-phosphate, α-glycerol phosphate, etc.) have low standard free energies of hydrolysis in the vicinity of −2 to −4 kcal/mole. Formation of these compounds by phosphorylation of the parent alcohol with ATP is thus very

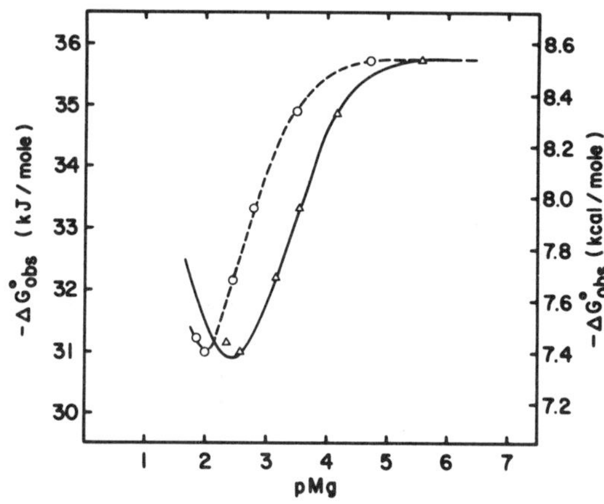

Figure 2.2. Free energy of hydrolysis of ATP at 38°C, pH 7.0, and ionic strength 0.25 as a function of free $[Mg^{2+}]$. The figure shows the observed standard free energy for the reaction

$$ATP + H_2O \rightleftarrows ADP + P_i$$

which was calculated from the combined equilibrium constant of the acetate kinase and phosphate acetyltransferase reactions and from the observed free energy for the hydrolysis of acetyl–CoA under these conditions (for details, see Guynn and Veech[7]). The symbols are: 0–0, experimental values plotted against the total Mg concentration; Δ–Δ, experimental values plotted against the calculated free $[Mg^{2+}]$. The curves over which the points are superimposed were calculated[7] for $[Mg^{2+}] = 0$. (Taken from Reference 7 and used with permission of the copyright holder.)

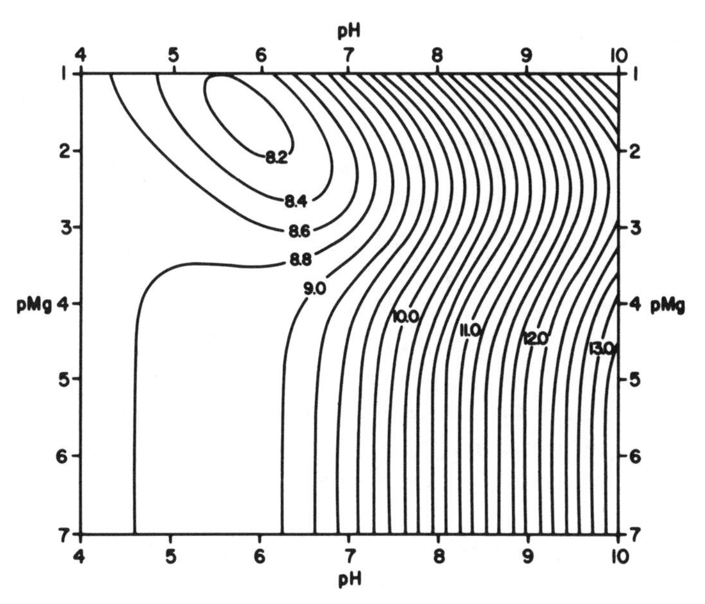

Figure 2.3. Contour map of $-\Delta G^{\circ\prime}$ in kilocalories per mole for the hydrolysis of ATP to ADP and P_i at 25°C and ionic strength 0.2. The contour lines are at intervals of 0.2 kcal/mole. (Taken from Reference 4 and used with permission of the copyright holder.)

Table 2.2. $\Delta G^{\circ\prime}$ for Hydrolysis of Phosphorylated Compounds

	$\Delta G^{\circ\prime}$	
Compound	kcal/mole	kJ/mole
Phosphoenol pyruvate	−14.8	−61.9
1,3-diphosphoglycerate (1-P)	−11.8	−49.3
Phosphocreatine	−10.3	−43.1
Acetyl phosphate	−10.1	−42.3
Phosphoarginine	−7.7	−32.2
ATP (→ADP + P_i)	7.6	−31.8
ATP (→AMP + PP_i)	−8.8	−37.4
Glucose-1-phosphate	−5.0	−20.9
Fructose-6-phosphate	−3.8	−15.9
Glucose-6-phosphate	−3.3	−13.8
Glycerol-1-phosphate	−2.2	−9.2

favorable thermodynamically, and these reactions may be considered to be physiologically irreversible.

2. Compounds that can be viewed as mixed acid anhydrides (1,3-diphosphoglycerate, acetyl phosphate, etc.) or substituted phosphoramidates (phosphocreatine, phosphoarginine, etc.) have much higher $\Delta G^{\circ\prime}$ values for hydrolysis ranging near -10 to -12 kcal/mole. Phosphoenolpyruvate is especially unstable thermodynamically, with a $\Delta G^{\circ\prime}$ for hydrolysis of -14.6 kcal/mole.

3. The $\Delta G^{\circ\prime}$ for hydrolysis of ATP (either the α–β or the β–γ linkage) is intermediate between these two groups. This is particularly opportune for metabolism, since it provides for the ability of ATP to serve as a common intermediary between the two groups. Thus "high-energy" compounds such as 1,3-diphosphoglycerate and phosphoenolpyruvate, produced by catabolic processes, can transfer their phosphoryl groups to form ATP, which, in turn, can act to phosphorylate compounds that are low on the scale of $\Delta G^{\circ\prime}$.

Structural Rationale for $\Delta G^{\circ\prime}$ for ATP Cleavage

Two major contributors to the large $\Delta G^{\circ\prime}$ for hydrolysis of ATP and other "high-energy" phosphate compounds are apparent, namely, charge effects and opportunity for resonance stabilization. Since the $\Delta G^{\circ\prime}$ for hydrolysis represents the difference in the free energy of the reactants and products of a reaction, we must consider the thermodynamic stability of ATP compared with that of hydrolytic products, ADP, and P_i. At neutral pH, ATP bears close to four negative charges that cause internal repulsion and thermodynamic instability. Cleavage to ADP and P_i (or to AMP and P_i) is accompanied by separation of negative charge, accounting in part for the large change in free energy. When ATP is complexed to a metal ion such as Mg^{2+}, the degree of internal charge repulsion is reduced, and this is reflected by the reduced $\Delta G^{\circ\prime}$ for hydrolysis of the Mg(II) complex of ATP (see Figure 2.3). By contrast, hydrolysis of a simple ester of phosphoric acid such as glucose-6-phosphate is not accompanied by equivalent charge separation, and the $\Delta G^{\circ\prime}$ is accordingly much lower. Furthermore, the polyphosphate portion of the ATP molecule is less stable than the hydrolysis products because the latter have greater opportunity for resonance delocalization of electrons. The bonds between the bridge oxygen and phosphorus atoms of ATP have essentially single-bond character, but after hydrolysis the bonds to these atoms can acquire one-third double-bond character by electron delocalization. The sum of the delocalization of electrons in ADP and P_i, compared to ATP, thus contributes further to the relatively large $\Delta G^{\circ\prime}$ for hydrolysis.

It might be expected that the thermodynamic instability of ATP could be reflected by uncharacteristic lengths and angles for the oxygen–phosphorus bonds. Unfortunately, investigation of this possibility by X-ray crystallography has been very difficult, mainly because of the high water content of crystals of disodium ATP and their instability and the large number of ATP molecules per unit cell of the crystals.[9] The most recent refinement of the crystal structure indicated that the β–γ bridge was the *shortest* of the three.[10] This result would not be expected on the basis of the simplest considerations and the fact that this bond is most commonly cleaved; it is quite possible, however, that the length of the β–γ bridge might be strongly influenced by coordination to divalent cations. It has been suggested[10] that one of the advantages gained by nature's use of bonds to phosphorus for energy storage could be the strong influence of metal coordination or bond distortion on the bond order. Thus slight distortion attributable to the orientation of the terminal phosphoryl group on the surface of an enzyme and the metal coordination dictated by the specific enzyme binding sites (see Chapter 4) might render the β–γ linkage more unstable (both thermodynamically and kinetically). In other words, compounds such as ATP may be particularly good targets for catalytic stresses.

COORDINATION OF METAL IONS

Divalent cations, especially Mg^{2+}, are strongly coordinated to the polyphosphates of ATP and other nucleotides. Table 2.3 lists some of the dissociation constants for various metal ions to ATP and some of its analogues, measured primarily by nuclear magnetic resonance (NMR) techniques.

There is some controversy over the site of coordination of metal ions to ATP, both in solution and bound to the active sites of enzymes. Although metal coordination has been found to be accompanied by a significant change in the ^{31}P-NMR chemical shift of ATP and its analogues, it has been clearly argued by Jaffe and Cohn[11] that it is hazardous to use the relative magnitudes of chemical shifts of the three phosphates for assigning the sites of coordination of metals. Nevertheless, many investigators have done so: for Mg–ATP with essentially the same ^{31}P-NMR chemical-shift data, Cohn and Hughes[12] proposed that magnesium was attached to the β- and γ-phosphoryls; Kuntz and Swift[13] suggested that the metal was coordinated to all three phosphoryls, and Tran-Dinh et al.[14] concluded that Mg was attached to the β position only. However, the coordination of magnesium to the γ phosphoryl is unequivocally established by the effect of

Table 2.3. Dissociation Constants for Metal Complexes of ATP and its Analogues (25°C)[a]

		K_d ($M \times 10^4$)		
Nucleotide	pH	Mg^{2+}	Ca^{2+}	Mn^{2+}
ATP	8.5	0.67	1.8	0.31
ATP	7.2	2.2	5.5	
AMPPNP[b]	8.5	0.26	0.85	0.12
AMPPCP	7.4	0.78	2.1	

[a]Data taken from R. G. Yount, D. Babcock, W. Ballantyne, and D. Ojala, *Biochemistry* **10,** 2484 (1971).
[b]Nonstandard abbreviations: AMPPNP, β,γ-imido analogue of ATP (adenylyl imidodiphosphate); AMPPCP, βγ-methylene analogue of ATP [adenosine 5′-(β,γ-methylene triphosphate].

the metal on the pK_a of this group—for ATP and its imido and methylene analogues, metal coordination causes a drop in the value of the pK_a by about 1.5 units (Table 2.4). Moreover, the lower K_d values for Mg^{2+} with ATP and its analogues compared to AMP and other nucleoside monophosphates suggest that the metal is coordinated to more than one phosphoryl group. (Since protonation of the γ but not the β phosphate can lead to a change of −4 ppm and the β chemical shift of AMP–PNP, it is questionable to conclude coordination of metal to β-P on the basis of shift data alone.) Similarly, the shift data cannot be used to define the role of α-P in the structure of Mg–ATP since the α-P resonance exhibits a small shift (ca. −0.3 ppm), of the same order as that of the γ-P at high pH. However, for Mg–ATP-αS (see Table 2.4), the α-P resonance is shifted about −2 ppm, but this cannot be taken as evidence for coordination to α-P because of the enhanced sensitivity of the internal thiophosphate residue to perturbations in the chemical environment. Because of the same kinds of consideration, chemical shift data cannot be taken alone as criteria to determine the site(s) of magnesium binding in enzyme-bound Mg–ATP.

Taken together, all the foregoing data and arguments support a structure in which the magnesium atom is closely associated with both the β and γ phosphates of Mg–ATP in solution. However, since the coupling constants $J_{\alpha\beta}$ and $J_{\beta\gamma}$ change similarly with metal binding for ATP and its analogues (Table 2.4), Vogel and Bridger[15] have proposed a model in which all possible mono-, bi-, and tridentate coordination structures exist in solution in fast equilibrium, with

Table 2.4. pK_a Values, NMR Parameters for ATP, ADP, and Their Analogues (±Mg)[a]

		NMR Parameters					
			Chemical Shifts, ppm[b]			Coupling Constants	
Compound	pK_a	pH	α-P	β-P	γ-P	$J_{\alpha\beta}$	$J_{\beta\gamma}$
ATP	6.7	8.0	−10.9	−21.3	−6.0	19.2	19.5
MgATP	5.3	8.0	−10.7	−19.1	−5.5	16.2	15.6
ADP	6.8	8.0	−9.7	−5.3		21.7	
MgADP	5.3	5.5	−10.2	−7.0		17.9	
AMPPNP[c]	7.7	9.0	−10.18	−6.43	−0.11	20.2	4.9
	6.2	9.0	−10.09	−5.33	−0.97	17.3	8.2
AMPPCP	8.0	11.0	−10.1	14.5	12.4	26.5	8.0
MgAMPPCP	6.2	10.0	−9.8	15.7	12.7		
AMPCP	8.1	11.0	23.0	12.0		9.0	
MgAMPCP	6.4	10.0	23.2	12.6			
AMPCPP	6.8	9.5	19.6	6.9	−5.4	8.5	25.0
MgAMPCPP	5.1	9.0	18.7	9.8	−4.9		
ADP-αS(A)[d]		7.3	42.1	6.3		30.8	
MgADP-αS(A)		7.6	44.7	6.4		28.0	
ADP-αS(B)		7.3	41.7	6.3		29.7	
MgADP-αS(B)		7.6	44.4	6.4		27.0	
ADP-βS	5.2	8.0	−11.1	33.9			
MgADP-βS	<4.0	8.0	−11.0	35.2		28.1	
ATP-αS(A)		8.0	43.4	−22.2	−5.5	27.5	20.5
MgATP-αS(A)		8.0	45.4	−20.6	−5.5	26.1	15.8
ATP-βS(A)		8.1	−11.4	29.9	6.0	26.5	27.3
MgATP-βS(A)		8.0	−11.2	36.3	5.7	28.1	27.5
ATP-βS(B)		8.1	−11.2	30.0	6.0	26.5	27.3
MgATP-βS(B)		8.0	−11.4	35.9	5.8	28.1	28.0
ATP-γS	5.3	8.4	−10.6	−22.0	35.0	19.6	29.0
MgATP-γS	<4.2	8.4	−10.6	−20.6	36.8	15.5	25.4
ATP-γF		8.0	−10.1	−22.6	−17.5		
MgATP-γF		8.0	−9.9	−22.4	−18.0		
ADP-βF		8.0	−10.8	−17.3		19.8	
MgADP-βF		8.0	−11.6	−17.6			

[a]Compiled from S. Tran-Dinh and M. Roux, *Eur. J. Biochem.* **76,** 245 (1977); E. Jaffe and M. Cohn, *Biochemistry* **17,** 652 (1978); H. J. Vogel and W. A. Bridger, *Biochemistry* **21,** 394 (1982).

[b]Referenced to external 85 percent H_3PO_4. Down-field shifts are given a positive sign.

[c]The nonstandard abbreviations are: AMPPNP, β,γ-imido analogue of ATP; AMPPCP, β,γ-methylene analogue of ATP; AMPCP, α,β-methylene analogue of ADP; AMPCPP, α,β-methylene analogue of ATP; ADP-αS, etc.: thiophosphoryl analogues of nucleotides, sulfur substitution on indicated phosphoryl group; ATP-γF and ADP-βF: fluoro analogues of ATP and ADP, fluorine substitution in terminal phosphoryl group.

[d]Symbols (A) and (B) refer to diastereoisomers as defined by F. Eckstein and R. S. Goody, *Biochemistry* **15,** 1685 (1976).

the interconversion between the different coordination structures faster than the Mg^{2+} on and off rates. Thus the ^{31}P-NMR spectrum would be the fast-exchange average of these structures. Similar suggestions were recently made for the structures of ATP and ADP.[16,17] Since the use of thiophosphate- and exchange-inert analogues has indicated that most enzymes show strong preferences for specific coordination structures, one can visualize a situation in which the different coordination structures are all in varying proportions and in fast equilibrium in solution, and an enzyme can select the one it specifically recognizes.[16,18,19] Unfortunately, however, this also means that the ^{31}P-NMR spectra of solution and enzyme-bound Mg–ATP and Mg–ADP cannot be directly compared, since this would require knowledge of the chemical shifts for all different coordination structures in solution, for resolution of contributions to the chemical shift caused by binding to the enzyme.

Another approach to determination of the structure of Mg–ATP in solution has been to study the changes in relaxation properties due to titration with paramagnetic metal ions.[12,20,21] The results of these investigations placed the metal ion closer to the β- and γ- than the α-P. Although this could be easily accommodated in the interpretations presented here by assuming that the equilibrium of all coordination structures is not random, it must be borne in mind that quite different binding constants have been observed for different metal ions (Table 2.3). It may thus be dangerous to extrapolate the results with paramagnetic metals to the Mg–ATP complexes.

REFERENCES

1. M. Cohn, in *NMR in Biochemistry,* S. J. Opella and P. Lu, Eds., Marcel Dekker, New York (1979), p. 7.
2. T. Benzinger, C. Kitzinger, R. Hems, and K. Burton, *Biochem. J.* **71,** 400 (1959).
3. L. Levintow and A. Meister, *J. Biol. Chem.* **209,** 265 (1954).
4. R. A. Alberty, *J. Biol. Chem.* **243,** 1337 (1968).
5. R. C. Phillips, P. George, and R. J. Rutman, *J. Biol. Chem.* **244,** 3330 (1969).
6. J. Rosing and E. C. Slater, *Biochim. Biophys. Acta* **267,** 275 (1972).
7. R. W. Guynn and R. L. Veech, *J. Biol. Chem.* **248,** 6966 (1973).
8. R. W. Guynn, H. J. Gelberg, and R. L. Veech, *J. Biol. Chem.* **248,** 6957 (1973).
9. O. Kennard, N. W. Isaacs, J. C. Coppola, A. J. Kirby, S. Warren, W. D. S. Motherall, D. G. Watson, D. L. Wamper, D. H. Chenery, A. C. Larson, K. A. Kerr, and L. Riva di Sanseverino, *Nature* **255,** 333 (1970).
10. K. A. Kerr, personal communication.

11. E. K. Jaffe and M. Cohn, *Biochemistry* **17,** 652 (1978).
12. M. Cohn and T. R. Hughes, *J. Biol. Chem.* **237,** 176 (1962).
13. G. P. P. Kuntz and T. J. Swift, *Federation Proc.* **32,** 546 (1973).
14. S. Tran-Dinh, M. Roux, and M. Ellenberger, *Nucleic Acids Res.* **2,** 1101 (1975).
15. H. J. Vogel and W. A. Bridger, *Biochemistry* **21,** 394 (1982).
16. W. W. Cleland and A. S. Mildvan, *Adv. Inorg. Biochem.* **1,** 163 (1979).
17. F. Ramirez and J. F. Mariceck, *Biochim. Biophys. Acta* **589,** 21 (1980).
18. A. S. Mildvan, *Adv. Enzymol. Relat. Areas Mol. Biol.* **49,** 103 (1979).
19. F. Eckstein, *Trends Biochem. Sci.* **5,** 157 (1980).
20. F. F. Brown, I. D. Campbell, R. Henson, C. W. J. Hirst and R. E. Richards, *Eur. J. Biochem.* **38,** 54 (1973).
21. P. Tanswell, J. F. Thornton, A. V. Korda, and R. J. P. Williams, *Eur. J. Biochem.* **57,** 135 (1975).

3

Mechanisms of ATP Production

Most processes that we regard as fundamental to the concept of life (cell growth and division, contractility, membrane polarization, etc.) consume energy in the form of ATP, with production of ADP and AMP. For maintenance of life, ADP and AMP are rephosphorylated to ATP by use of energy derived from the processes of catabolism. Adenosine triphosphate is generated in two important ways, namely by so-called substrate-level phosphorylations and by phosphorylations coupled to electron flow such as mitochondrial oxidative phosphorylation and photophosphorylation in chloroplasts. This chapter considers the detailed mechanisms of the enzymatic replenishment of ATP stores by these different systems.

SUBSTRATE-LEVEL PHOSPHORYLATIONS

In catabolism of glucose by means of glycolysis and the tricarboxylate cycle, there are three sites at which an equivalent of one ATP molecule is generated. These are by the combined action of glyceraldehyde 3-phosphate dehydrogenase and phosphoglycerate kinase, by pyruvate kinase (in the glycolytic pathway), and by succinyl–CoA synthetase in the TCA cycle. Some of the principles, if not the details, of the mechanisms of substrate-level phosphorylation may be relevant to enzymatic synthesis of ATP that is coupled to electron flow. Partly for this reason, the enzymes that catalyze substrate-level phosphorylations have received a great deal of attention, and some of the pertinent results are important to any comprehensive consideration of ATP synthesis. The modes of action of these enzymes, together with aspects of their structure that contribute to their function, are considered in the following.

Glyceraldehyde 3-Phosphate Dehydrogenase

The reaction catalyzed by this enzyme might be considered to be a substrate-level oxidative phosphorylation: a high-energy phosphate anhydride bond is formed, coupled to the oxidation of the aldehyde moiety of the glyceraldehyde 3-phosphate to the level of a carboxylate:

$$\begin{array}{l} CHO \\ | \\ CHOH \\ | \\ CH_2OPO_3^{2-} \end{array} + NAD^+ + P_i \rightleftharpoons \begin{array}{l} \overset{O}{\overset{\|}{C}}-O-PO_3^{2-} \\ | \\ CHOH \\ | \\ CH_2OPO_3^{2-} \end{array} + NADH + H^+$$

A comprehensive review of important aspects of the enzymology of this reaction appeared in 1976.[1]

As can be seen, later the mechanism of action of glyceraldehyde 3-phosphate dehydrogenase is intimately associated with interactions between its subunits. The enzyme is a tetramer of four chemically identical subunits. The amino acid sequences of the enzyme from such diverse sources as yeast, lobster, and pig muscle have been determined. The subunits from each of these have molecular weights of 35,000 and consist of about 330 amino acids, with remarkable conservation of sequence between the three. Two very significant structural features have been recognized for decades: the presence of a catalytically important –SH group (now known to be Cys-149) and the presence of a tightly bound coenzyme (NAD^+) in the enzyme as it has been isolated, with the latter contributing to the "Racker band" of ultraviolet absorbance at 330 nm.[2] The involvement of these features in catalysis is represented in the following steps:

$$E\begin{smallmatrix}\cdot NAD^+ \\ -SH\end{smallmatrix} \xrightarrow{RCHO} E\begin{smallmatrix}\cdot NAD^+ \\ -S-\underset{OH}{\underset{|}{C}}H-R\end{smallmatrix} \rightarrow E\begin{smallmatrix}\cdot NADH \\ -S-\underset{O}{\underset{\|}{C}}-R\end{smallmatrix} \xrightarrow{P_i} E\begin{smallmatrix}\cdot NADH \\ -SH\end{smallmatrix} + R-\overset{O}{\overset{\|}{C}}-O-PO_3^{2-}$$

There is a wealth of information that suggests cooperative behavior between subunits during the above reactions. For example, MacQuarrie and Bernhard[3] took advantage of the "half-sites" behavior of the enzyme to prepare a derivative of glyceraldehyde 3-phosphate dehydrogenase in which two of the four apparently equivalent Cys-149 residues were blocked by catalytic thioester formation using the substrate analogue furylacryloyl phosphate and in which the two remaining Cys-149 sulfhydryls were alkylated by reaction with iodoacetamide. Furylacryloyl groups were then removed by arsenolysis, and the resulting dialkylated enzyme was tested for stoichiometry of thioesterification by subsequent reaction with excess furylacryloyl phosphate: only one acyl group could be incorporated into the dialkylated enzyme, despite the fact that the arsenolysis step had just previously removed two acyl groups! These and other results are not easily reconciled with a model for half-sites behavior that involves induced assymetry but indicate that the enzyme may have preexisting assymetry that might better be represented by the formula $(\alpha\alpha_1)_2$, where α and α_1 represent chemically identical but conformationally distinguishable subunits, capable of interconversion during catalysis.

This concept has been given strong support from the results of high-resolution X-ray analysis of the muscle enzyme.[1,4,5] Four entirely equivalent subunits, arranged tetrahedrally, would be expected to give rise to crystallographic 222 symmetry (i.e., having three orthogonal twofold axes). It has been demonstrated that the crystal structure approximates this situation, but that two of the axes are pseudo-twofold. This has important implications for the subunit structure since it suggests that two of the subunits are conformationally distinguishable from the other two. The identical pairs of subunits have been designated *red–yellow* and *green–blue*. Differences in the relative positions of atoms in the two pairs average only about 0.3 Å, but certain individual amino acid residues differ by as much as 3 Å between the two conformations of subunits. The largest differences in the orientations of individual amino acids are in the region of NAD^+ binding and at the P_i site. For example, in the red–yellow subunits, Cys-149 is connected to NAD^+ by continuous electron density (accounting for the Racker band), whereas in the green–blue pair, the crucial –SH group appears to be associated with His-176. Furthermore, in the latter pair, a loop of protein has moved 2 Å closer to the coenzyme, thereby generating a site for P_i binding that does not seem to exist in the former. Other important differences include the orientation of the adenine ring of the coenzyme in the two structures and in the interaction of its ribose moieties with the protein, perhaps accounting for anti-cooperative binding of NAD^+ to the enzyme. These kinds of difference in the active site suggest that the mechanism of catalysis may involve a flip-flop in the conformations of individual subunits coupled to specific partial reactions: catalysis may begin with the red–yellow form and involve conversion of the site to the green–blue configuration by the time that thioester bond formation has occurred.[4] The possibility of such alternating-sites behavior is perhaps crucial for effective catalysis, and it is of particular interest in the context of this discussion that alternating-sites cooperativity has been suggested for catalysis by other enzymes that produce ATP directly, namely, electron-transport-linked ATP synthetases and succinyl–CoA synthetase (see below).

Phosphoglycerate Kinase

The reaction catalyzed by this enzyme involves the transfer of a phosphoryl group from the carboxyl moiety of 1,3-diphosphoglycerate to produce an equivalent of ATP:

$$\begin{array}{l} \mathrm{O} \\ \| \\ \mathrm{C{-}OPO_3^{2-}} \\ | \\ \mathrm{CHOH} \\ | \\ \mathrm{CH_2OPO_3^{2-}} \end{array} + \mathrm{ADP} \overset{\mathrm{Mg^{2+}}}{\rightleftharpoons} \begin{array}{l} \mathrm{COO^-} \\ | \\ \mathrm{CHOH} \\ | \\ \mathrm{CH_2OPO_3^{2-}} \end{array} + \mathrm{ATP}$$

This accounts for one-half of the ATP that is generated by anaerobic glycolysis.

Phosphoglycerate kinase is present in anomalously high concentrations in vivo.[6] For example, in muscle the concentration of this enzyme is approximately 2–5 mg/g wet weight, or 0.1 m*M*. Thus its concentration is roughly equal to that of one of its substrates (ADP) and an order of magnitude greater than the free concentration of the other (1,3-diphosphoglycerate). It has been estimated that there is a 25-fold excess of the activity of phosphoglycerate kinase, relative to the maximum rate of glycolysis in muscle.

Molecular Properties

Phosphoglycerate kinase from a variety of sources (yeast, muscle, erythrocyte, etc.) has been found to be composed of a single subunit of molecular weight near 45,000.[6] Preliminary X-ray analysis of the structures of the enzyme from yeast[7,8] and horse muscle[9,10] have shown remarkable similarity in tertiary structure. The single polypeptide is folded into two domains of approximately equal size. X-Ray crystallography at higher resolution[11] indicates that the metal complexes of ADP and ATP are bound to one of the domains in a region of the molecule that cannot accommodate the binding of phosphoglycerate. It is proposed that the binding site for the latter substrate is located on the other domain about 10 Å from the nucleotide binding site. This would thus require the operation of a major hinge mechanism at the neck of the molecule that connects these two domains. Bending the hinge by about 10–20° would thus bring the two substrates together in space and would expel solvent to allow for catalysis in a water-free environment. Working independently on yeast phosphoglycerate kinase by use of small-angle X-ray-scattering techniques, Pickover et al.[12] have shown that the formation of the ternary enzyme–MgATP–3-phosphoglycerate complex is accompanied by a reduction in the radius of gyration of the enzyme. Similarly, these workers interpreted these results in terms of a hinge motion that brings the two domains of the enzyme together.

Mechanism of Phosphoryl Transfer

There has been some controversy over the mechanism by which the phosphoryl group is transferred during catalysis by phosphoglycerate kinase, particularly concerning the possible involvement of a phosphorylated enzyme intermediate. For example, it has been reported[13,14] that the erythrocyte enzyme could be converted to an acid- and base-labile and hydroxylamine-sensitive phosphorylated derivative, perhaps containing a phosphorylated glutamyl side chain. Strong doubts were cast on such a phosphorylated intermediate by the demonstration that the purified enzyme can contain close to one molar equivalent of tightly bound 1,3-diphosphoglycerate,[15] perhaps accounting for the results that had earlier suggested a phosphoenzyme intermediate.

Since single steps involving phosphoryl transfer are known to proceed by means of in-line attack with net inversion of stereochemical configuration,[16] a mechanism without the involvement of a transient phosphoenzyme intermediate would be expected to be accompanied by inversion. This has been confirmed by the elegant study by Webb and Trentham,[17] who prepared chiral ATP-γS containing distinct stereochemical configurations of ^{16}O, ^{17}O, and ^{18}O. The transfer of the thiophosphoryl group to the acceptor 3-phosphoglycerate was shown to occur with net inversion of configuration, lending very strong support to the proposal that phosphoryl transfer catalyzed by this enzyme is direct, without the transient formation of a phosphoenzyme form by a double-displacement reaction that would have been expected to result in overall retention of stereochemical configuration.

One feature of catalysis by phosphoglycerate kinase may be particularly illustrative of a property of enzymes that produce ATP and many that consume it. This feature is the effect of the enzyme on the position of equilibrium of the enzyme-bound reactants. The equilibrium constant for the phosphoglycerate kinase reaction is approximately equal to 3000 in the direction of ATP synthesis, and the presence of catalytic quantities of the enzyme, of course, has no effect on this value. However, by measuring the concentrations of enzyme-bound reactants by ^{31}P–NMR at high enzyme levels, Nageswara Rao et al.[18] have shown that the equilibrium constant for the interconversion of the bound reactants is very close to unity. Although this may seem at first to represent a thermodynamic impossibility, it was first recognized by Haldane[19] and later by others[20,21] that the position of equilibrium of enzyme-bound reactants is a function of their respective dissociation constants; stated simply, if a reactant is very tightly bound, this will tend to favor its production on the surface of the enzyme. As is seen below in the consideration of the electron transport-linked ATP synthesis, this may be a very crucial principle adopted in other ATP-producing systems. In this context, one might recall that the concentration of phosphoglycerate kinase *in vivo* is comparable to that of its substrates and that most of the 1,3-diphosphoglycerate in the cytosol exists as its complex with this enzyme. Thus the equilibrium constant for enzyme-bound reactants may be of physiological significance.

Pyruvate Kinase

Pyruvate kinase catalyzes the reaction

$$\text{Phosphoenolpyruvate} + \text{ADP} \rightleftharpoons \text{Pyruvate} + \text{ATP}$$

The enzyme is unusual in its cation requirements: both a divalent cation (normally Mg^{2+}) and a monovalent cation (normally K^{+}) are required for activity.[21] The steps of the reaction are believed to include the transfer of the phosphoryl group from phosphoenolpyruvate to ADP to form ATP and enolpyruvate, followed by quasiequilibrium proton donation to produce the keto tautomer of pyruvate.[22] The observation that pyruvate kinase catalyzes the rapid detritiation of [^{3}H]phosphoenolpyruvate in the presence of ADP indicates that the release of the product pyruvate may be the slowest step in the overall reaction.[22] It is most intriguing that the equilibrium constant for the interconversion of the pyruvate kinase reactants is affected by their binding to the enzyme in a fashion similar to that described above for phosphoglycerate kinase. Here again, the equilibrium constant is shifted by about three orders of magnitude to near unity for the transfer of the phosphoryl group from the donor to the acceptor on the surface of the enzyme.[23]

Nucleotide Binding Site

By a variety of paramagnetic probes and NMR, considerable information has been made available about the geometry of the active site of pyruvate kinase, particularly the arrangement of the substrates in relation to the metal ions that are required for activity. Thus two divalent metal ions have been shown to be associated with each active site.[24] One of these is coordinated to the enzyme and forms second-sphere complexes with the bound substrates. The other divalent cation has been shown to be coordinated directly to the phosphate of ATP, probably by way of the γ- and β-phosphoryl groups. Further information independent of NMR measurements has been provided by the use of exchange-inert chromium complexes of ATP.[25] Four tridentate Cr(III)–ATP complexes and stereochemical isomers of the β,γ bidentate complex were prepared and tested as substrates and promoters of partial reactions. Only the Δ isomer of the β,γ-bidentate complex showed significant activity as a substrate for phosphoryl transfer, and this isomer was most effective in promoting the enolization of pyruvate catalyzed by this enzyme, although all forms of Cr–ATP had significant effect on the latter reaction. Figure 3.1 shows a proposal for the mechanism of action of pyruvate kinase that is consistent with these results and with the NMR data, showing likely positions of the two divalent cations and single monovalent cation relative to the substrates.

The results of high-resolution X-ray crystallographic analysis of pyruvate kinase from cat muscle[26] have indicated the presence of two nucleotide binding sites on this enzyme, one in a cleft between two domains (A and C) of the molecule and one at the active site buried entirely within domain A. This idea has been strengthened by the kinetics of inactivation of pyruvate kinase by 5′*p*-fluorosulfonylbenzoyl adenosine, an affinity label designed by Colman and her colleagues to be specific for ATP binding sites.[27,28] Annamalai and Colman[28] correlated fast reaction of this reagent with a cysteine residue at the cleft (non-catalytic) ATP binding site. The active site contains two amino acid side chains

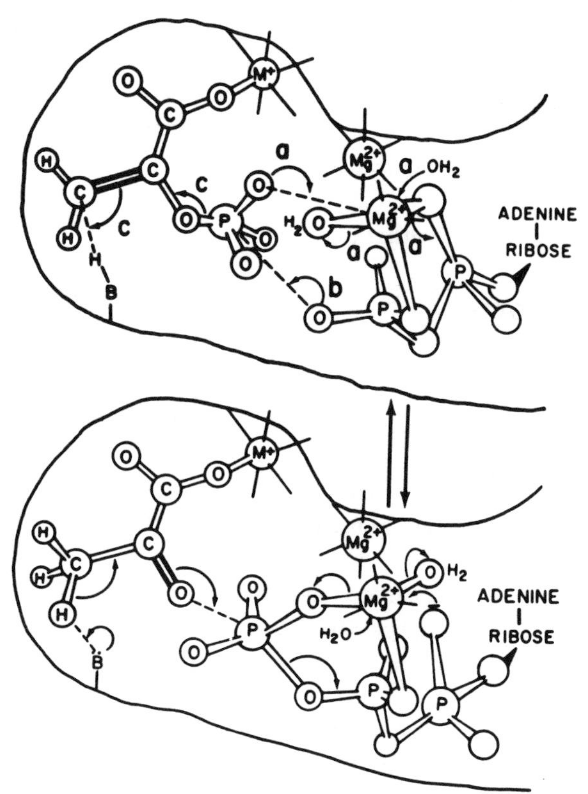

Figure 3.1. Mechanism proposed[25] for the mechanism of action of pyruvate kinase. In step a, Mg(II) migrates from α,β-ADP coordination to β-ADP, phosphoenolpyruvate coordination by two ligand substitution reactions that may be separate or concerted. In step b, the phosphoryl group is transferred, yielding the delta isomer of the β-γ-bidentate MgATP complex and the enolate of pyruvate. In step c, the enolate of pyruvate is protonated and converted to the keto tautomer.

that are attacked by the reagent, that of tyrosine and of cysteine. (The latter side chain reacts slowly relative to the cysteine at the noncatalytic site). The pattern of protection of the enzyme from this affinity labeling reagent by metal ions and substrates is in harmony with the proposed structure of the active site depicted in Figure 3.1: MgADP protects almost completely against tyrosine modification, whereas free Mg^{2+} ion protects against modification of cysteine at the catalytic site. These results thus confirm the concept of two divalent cation binding subsites at the active center of pyruvate kinase.

The mechanism depicted in Figure 3.1 indicates direct transfer of the phosphoryl group from phosphoenolpyruvate to ADP without the intermediate involvement of a covalent phosphorylated enzyme derivative. Such direct transfer has been confirmed by the elegant stereochemical experiments by Knowles and his colleagues,[29,30] who showed that the pyruvate kinase reaction is accompanied by inversion of the stereochemical configuration of [γ–(S)–^{16}O, ^{17}O, ^{18}O]-ATP.

Succinyl–Coenzyme A Synthetase

This enzyme catalyzes the substrate-level phosphorylation that is a component of the tricarboxylate cycle:

$$\text{Succinyl–CoA} + \text{NDP} + P_i \overset{Mg^{2+}}{\rightleftharpoons} \text{Succinate} + \text{CoA} + \text{NTP}$$

The nucleotide specificity is interesting and somewhat enigmatic; for example, the enzyme from Gram-negative bacteria and plant mitochondria phosphorylates ADP to produce ATP directly, whereas the mammalian mitochondrial enzyme is specific for guanine nucleotides and produces ATP only by the subsequent action of nucleoside diphosphokinase.[31] As representatives of these two groups, succinyl–CoA synthetases from *Escherichia coli* and from pig heart have received by far the most attention.

In contrast to the other enzymes that catalyze substrate-level phosphorylations (see above), interest in succinyl–CoA synthetase has been sparked in part by the covalent phosphorylated intermediates that have been proposed to participate in its catalytic mechanism. For the *E. coli* enzyme, these may be formulated:[31]

$$\text{E} + \text{ATP} \rightleftharpoons \text{E–PO}_3 + \text{ADP}$$
$$\text{E–PO}_3 + \text{Succinate} \rightleftharpoons \text{E} \cdot \text{succinyl–PO}_3$$
$$\text{E} \cdot \text{succinyl–PO}_3 + \text{CoA} \rightleftharpoons \text{E} + \text{Succinyl–CoA} + P_i$$

The phosphoenzyme form (E–PO_3) is known to contain a 3-phosphohistidyl residue[31] and to have the kinetic behavior expected for an obligatory catalytic intermediate:[32] E–PO_3 is formed at a rate exceeding the steady-state rate of the overall reaction, and the steady-state level of this form is attained before the overall reaction enters its steady state.* The second intermediate proposed, tightly

*Considerable excitement was generated in the early 1960s by the discovery of protein-bound phosphohistidine in mitochondria[33] because its properties were those that might have been expected for a covalent X–PO_3 intermediate in oxidative phosphorylation. The phosphohistidine under investigation was subsequently relegated to the E–PO_3 form of succinyl–CoA synthetase.[34]

bound succinyl phosphate,[35] has yet to be shown to meet these kinetic criteria but does account nicely for the early experiments showing that one of the oxygen atoms in the P_i product originated from succinate.[36]

Involvement of Enzyme Subunits in Catalysis

Succinyl–CoA synthetase contains two types of subunits. For the *E. coli* enzyme, the α and the β subunits have molecular weights of approximately 29,500 and 38,500, respectively,[37] and these are slightly but significantly larger in the pig heart enzyme.[38] The mammalian enzyme exists predominantly as an αβ dimer, whereas the bacterial enzyme strongly favors an $\alpha_2\beta_2$ "dimer of dimers" structure. In both cases the site of enzyme phosphorylation is the α subunit; significantly, in the tetrameric *E. coli* enzyme only one of the two α subunits is phosphorylated by excess ATP.[39,40]

It has been suggested that *E. coli* succinyl–CoA synthetase catalysis may involve cooperative interactions between its subunits to promote specific partial reactions. This was first suggested by the elegant experiments by Bild et al.,[41] who showed that the relative rate of exchange of ^{18}O between phosphate and succinate is a strong function of the concentration of ATP. This was interpreted (for the reaction in the direction of succinyl–CoA synthesis) in terms of ATP binding at one site promoting a step involving succinyl phosphate production or succinyl–CoA release at the adjacent site on the neighboring half of the enzyme molecule. Capacity for alternating-sites cooperativity was subsequently confirmed by Wolodko et al.,[42] who showed that both α subunits in an individual $\alpha_2\beta_2$ enzyme molecule participate in the reaction by undergoing transient phosphorylation. It now appears that the step of the overall reaction that is promoted by ATP binding at the neighboring site may be the one involving the synthesis of enzyme-bound succinyl phosphate, since it has been shown by ^{31}P-NMR that no detectable succinyl phosphate is produced by a mixture of E–P and CoA until ATP is subsequently added.[43,44] Such cooperative behavior of subunits in promoting catalytic steps has been independently proposed for the electron-transport-linked F_0F_1 and CF_0CF_1 ATPases (see below).

ELECTRON TRANSPORT-LINKED ATP SYNTHESIS

There are few areas of biochemical research that have attracted so much attention and resulted in so many elegant and divergent theories over the past 30 years as the synthesis of ATP that is coupled to electron flow. In respiring mitochondria, the actual step involving ATP synthesis is catalyzed by the F_0F_1-ATPase. The F_0 component of this system is a constituent of the mitochondrial inner membrane and is responsible for coupling ATP synthesis to electron transport by mechanisms to be discussed in more detail below. The F_1 component is a soluble major fragment of the mitochondrial membrane ATP-synthesizing system, which—when free in solution and not associated with F_0—catalyzes the hydrolysis of ATP. In bacterial systems, the corresponding component ATPase is sometimes referred to as the BF_1–ATPase, and in the photosynthetic system of chloroplasts, the production of ATP is catalyzed by the CF_1–ATPase in association with the CF_0 component. Although these three ATP-producing enzymes may differ in the details of their structure and mechanism of action, it seems clear that the principles of their catalytic mechanisms and their association with their respective electron transport systems are close to equivalent. For this reason, they are considered here without differentiation except where appropriate.

The membranes that enclose mitochondria, bacteria, and chloroplasts are extraordinarily complex in structure. As integral components, they contain a system of catalysts for the coupling of the energy derived from exergonic electron flow to the synthesis of ATP from ADP and inorganic phosphate. Early ideas about the mechanism of these electron transport-linked reactions were focused on the roles of the electron carriers in the energy-conserving process. It was variously suggested that a high-energy form of the redox component, either bound to a specific ligand such as P_i or in a specific conformation, was the link between electron transport, ATP production, and other oxidation–reduction-dependent properties (e.g., ion transport) of this specialized membrane.[45] These concepts have been replaced by the Nobel prize-winning chemiosmotic hypothesis proposed by Peter Mitchell.[46] The essence of his hypothesis for energy capture from electron transport is that the electron carriers are so arranged in the membrane that exergonic electron transport is coupled with the transport of protons across the membrane (from inside to outside in mitochondria and bacterial cells and from outside to inside in chloroplasts and inverted submitochondrial particles). Electron flow thereby results in a protonmotive electrochemical gradient across the membrane, which may be relieved by the transport of protons back to their starting side through a proton channel in the membrane (the hydrophobic F_0 component), which is connected to the F_1–ATPase. Thus a proton circuit is constructed to couple the exergonic reaction (respiration or photochemically driven electron flow) to the endergonic synthesis of ATP from ADP and P_i by the F_1–ATPase. Although Mitchell's hypothesis was certainly not universally well received for several years after it was enunciated, there is no longer any question that proton pumps are present in these specialized membranes to connect electron flow with ATP production. Thorough review of the evidence in support of the chemiosmotic theory is far beyond the scope of this work, as is a complete discussion of the mechanism of energy capture and transmission in the membrane in the form of electrochemical gradient. For these, the reader is referred to several comprehensive reviews.[45,47–51]

Structure of F_1–ATPase

The structure of the F_1–ATPase has been probed by a variety of techniques. In the early 1970s, electron microscopy of the mitochondrial membrane indicated that the ATPase was a relatively large globular protein, attached to the F_0-membrane component perhaps by means of a stalk structure.[52] The F_1 component can be solubilized by mild treatment and purified in soluble form; here it catalyzes ATP cleavage, but when it is reassembled by plugging it back into the proton pore, it can regain the capacity for catalysis of chemiosmotically driven ATP synthesis.

Regardless of its pro- or eukaryotic source, the F_1–ATPase has been found to have a molecular weight of about 360,000 and has been found to contain five distinct subunit types. These subunits (with their approximate molecular weights) are designated α (55,000), β (50,000), γ (32,000), δ (18,000), and ε (8–15,000?). Although there is not complete agreement on the stoichiometry of association of the subunits, the most favored model for the F_1–ATPase involves three copies of each α and β subunit surrounding a single γ subunit, together comprising the head of the ATPase, and a stalk containing one each of the δ and ε subunits.[52,53] In addition to these constitutive subunits, mitochondrial F_1–ATPase has been shown to have a sixth polypeptide more loosely associated with it, namely, a specific inhibitor protein of molecular weight near 10,000.[54] This inhibitor is readily removed by mild treatment (heat, high pH, or gel filtration). It is believed that the inhibitor polypeptide binds most tightly *in vivo* when the ATP:ADP ratio is high and that it tends to dissociate during the high electron flux or when the ATP:ADP ratio drops.[58] The inhibitor is thought to prevent ATP hydrolysis when the energy state of the mitochondrion is low, but not to prevent synthesis of ATP when electron flow is high.

We now have ideas of the functional roles for each of the constitutive subunits of F_1–ATPase. Much of these have arisen from studies of the bacterial system, which is amenable to genetic manipulation. For example, several studies on the bacterial F_1–ATPase suggest that the δ subunit is involved in the attachment of the enzyme to the membrane. It has been found that *E. coli* mutants lacking only this subunit produce a fully active F_1–ATPase that cannot become associated with the membrane.[56–58] Addition of purified δ subunit to the mutant enzyme restores its ability to associate with membrane vesicles, also restoring the capacity for coupled ATP synthesis.[59]

The ε subunit may also be required for binding of F_1 to depleted membranes. This concept is strongly supported by the elegant experiments by Sternweis,[60] who prepared enzyme devoid of the ε subunit by immunoabsorbance chromatography. The addition of purified ε subunit was required to restore the ability of this derivative to bind to the membrane. Furthermore, Yoshida et al.[61] have shown that the ε subunit of the F_1–ATPase of a thermophilic bacterium binds specifically to the F_0 component that has been reconstituted into liposomes. There is no evidence that the ε subunit from mitochondrial F_1–ATPase plays such a role. This subunit appears to have a second function beyond this architectural one, however; it has been shown to be an inhibitor of soluble bacterial F_1–ATPase activity,[62] although it has no effect on the ATPase activity or energy-coupling properties of membrane-bound F_1, and the native enzyme is fully active even though it contains a copy of this subunit. It has been suggested[62] that the purpose of the inhibitory effects may be to prevent useless hydrolytic effects of the newly synthesized soluble enzyme before it becomes associated with the membrane or that it may mediate inhibition by some other yet uncharacterized component. Experiments of Sternweis (see Dunn and Heppel[53]) showing that antiserum raised against the ε subunit has an inhibitory effect on the membrane-associated enzyme suggest that interaction of this subunit with a putative protein modulator may be possible. The relationship of ε of the bacterial system to the F_1 inhibitor peptide of mitochondrial ATPase is still unclear.

The remaining three subunits comprise the $\alpha_3\beta_3\gamma$ head of the F_1 molecule. Each has been purified to homogeneity, allowing for extensive characterization of their individual properties and evaluation of their contributions to the overall function of the enzyme.

Although the catalytic machinery for the actual synthesis of ATP is thought to lie elsewhere (see below), the α subunit has been shown to bind nucleotides extremely tightly (K_d for ATP = 10^{-7} *M*).[63] The first-order rate constant for dissociation of ATP from its complex with α is 0.21 min^{-1}, some six orders of magnitude lower than the rates of dissociation of typical enzyme–substrate complexes.[64] This suggests that α may contain the site for the association of tightly bound ATP in the native enzyme that may be implicated in ATP synthesis, as is discussed below. The α subunit undergoes a major conformational change when ATP is bound, sufficient to be detectable by an increase in its sedimentation coefficient[64] or by inelastic light-scattering techniques.[65] One indication that the α subunit may play a role in catalysis is derived from the work of Bragg and Hou on the F_1–ATPase from a mutant of *E. coli* that is deficient in oxidative phosphorylation.[66] Although the mutant enzyme contains an altered α subunit, it can still interact with F_0 to reduce proton flow, but the enzyme has little ATPase activity; tightly bound ATP is associated with the mutant enzyme in normal amounts. These results may be better understood in the context of the proposed mechanism for ATP synthesis that involves catalytic turnover of tightly bound nucleotide (see below).

The β subunit is believed to contribute directly to catalysis of the formation of ATP from ADP and P_i. Much of the evidence in favor of a catalytic role for β is derived from chemical modification studies, particularly those involving reaction of the enzyme with photoaffinity labels and with dicyclohexylcarbodiimide.[52,53] Stoichiometric modification of a single carboxyl group in β by dicyclohexylcarbodiimide is accompanied by inactivation of the ATPase and loss of capacity for low-affinity nucleotide binding. In keeping with its high affinity for ATP, photoaffinity analogues of ATP preferentially label α, but significant label is also found to be associated with β, consistent with the idea that it possesses a site equipped for catalysis of ATP synthesis or hydrolysis.

The requirement of the γ subunit for reconstitution of catalytically active enzyme suggests that its role may be primarily architectural.[53] Strong interaction of γ with δ has been demonstrated, whereas neither isolated α nor β subunits interact with ε. Thus it is suggested that γ holds the $\alpha_3\beta_3$ head portion of the molecule to δ. All the foregoing is consistent with the results of the incisive chemical cross-linking studies by Bragg and Hou.[67] In addition to such a structural role, it has been suggested that the γ subunit from both bacterial[61] and chloroplast[68] F^1–ATPases may act as a proton gate. For example, systems containing chemically modified γ show increased leakage of protons without concomitant ATP production.[68]

Mechanism of ATP Synthesis by F_1–ATPase

There is now a clear consensus that protonmotive force, essentially as enunciated in Mitchell's chemiosmotic hypothesis, represents the form in which free energy is captured by electron transport in respiring mitochondria and bacteria and in chloroplasts. This, however, by no means solves the problems of covalent chemistry that are associated with the transformation of this membrane potential into chemical bond energy in the form of ATP. It seems clear that the proton gate to relieve the chemiosmotic potential is by way of the F_0F_1 system; many properties of these coupling factors that are reviewed above lead directly to this conclusion. In Mitchell's early formulations of his hypothesis,[46] it was proposed that the emerging protons play a direct role in catalysis of ATP synthesis on F_1. In his model, two protons flow down an electrochemical gradient through F_0F_1, to be directed toward the $H_2PO_4^-$ site on F_1 where they protonate and remove an oxygen atom to produce water and $H_2PO_3^+$. Many have felt, however, that mechanisms involving direct protonation of substrate are not terribly attractive and that evidence in their support is not compelling. For these reasons, many investigators continue attempts to identify steps in the process leading to ATP synthesis by the F_0F_1 system and the means by which the proton efflux is linked to this exergonic reaction.* Review of the current literature indicates that most of the experimental evidence is compatible with a mechanism of ATP synthesis involving the following:

1. Direct attack of ADP on P_i to form ATP, without the involvement of a covalent phosphorylated intermediate.
2. Production of *tightly bound* ATP from ADP and P_i without input of energy from proton translocation. The energy-requiring step in synthesis of free ATP seems to be the release of this product from the ATPase, and the energy may be used to drive conformational changes in the ATPase to allow the release of ATP.
3. Alternating site cooperativity or catalytic cooperativity between individual active sites on the $\alpha_3\beta_3$ portion of the ATPase system.

Each of these aspects of ATP synthesis is discussed more fully in the following.

Absence of Covalent Intermediates. With early precedents such as the covalent intermediate of glyceraldehyde 3-phosphate dehydrogenase (see above), it is easy to formulate a possible mechanism for ATP synthesis involving intermediates such as $X–PO_3$ or Y–ADP and so on. These ideas formed the basis for the chemical coupling hypothesis that was the subject of much investigation

*Some of the controversy that has surrounded alternate proposals to that of Mitchell for the mechanism of ATP synthesis has arisen from the misunderstanding that advocates of these proposals did not agree with other aspects of the entire chemiosmotic hypothesis. This is generally not true—the issue is simply to understand how the protonmotive gradient is translated into covalent chemistry.

in the years following its proposal by Slater.[69] Identification of such intermediates was sought by the use of radioactively labeled phosphate or ADP in systems catalyzing oxidative phosphorylation. Although there were many flurries of excitement when compounds such as phosphohistidine were thus discovered,[33] none of these have withstood scrutiny. Of course, failure to detect a covalent intermediate by no means constitutes proof that one does not exist. Recently, however, Webb et al.[70] have provided evidence that, taken together with the absence of a demonstrable intermediate, strongly indicates that the synthesis of ATP occurs by way of direct nucleophilic attack of ADP on Pi.* Using beef heart mitochondrial F_1–ATPase, these workers have studied the hydrolysis of ATP-γS, stereospecifically labeled with ^{18}O in the γ position, in ^{17}O–water. The product inorganic [^{16}O, ^{17}O, ^{18}O]thiophosphate is chiral, and analysis of the configuration of the product showed that the hydrolysis reaction was accompanied by stereochemical inversion. In view of the persuasive arguments by Knowles[16] that the phosphoryl transfer occurs by means of direct, in-line attack with stereochemical inversion, the observations of F_1–ATPase are clearly compatible with a single step in the transfer of the phosphoryl group from ADP to water and provide all the further evidence that was necessary to abandon the idea of a covalent phosphoenzyme intermediate in this process.

Tightly Bound ATP. At the International Congress of Biochemistry in Stockholm in 1973, Paul Boyer presented data and ideas suggesting that the point of energy input from proton translocation need not be the step involving formation of the covalent bond between the β and γ phosphoryls of ATP. One principal argument is derived from the fact that the exchange of ^{18}O between water and ATP remains rapid even in the presence of sufficient uncoupler concentration to block net ATP production.[72,73] This establishes that dynamic reversal of ATP synthesis can occur at the active site without energy input. This concept is further supported by the observations that F_1-catalyzed hydrolysis of very low concentrations of ATP is accompanied by extensive exchange of phosphate oxygen atoms with medium water oxygens, again demonstrating that dynamic reversal of ATP hydrolysis occurs. Independently, Slater and his colleagues[45] proposed that the energy-requiring step of ATP synthesis was the release of ATP. This suggestion was based on measurements of the ratio of bound ADP to bound ATP on isolated F_1 preparations that indicated that the equilibrium constant for

*Boyer[71] has shown that the β,γ bridge oxygen of ATP is provided by ADP, indicating that ADP oxygen attacks the phosphorus of P_i rather than the other way around.

ATP synthesis from ADP and P_i on the surface of the enzyme is near unity. There is nothing heretical about this; recall that equivalent measurements have shown changes of three orders of magnitude in the equilibrium constant for enzyme-bound reactants on phosphoglycerate kinase[18] and pyruvate kinase.[23] The relationship of this tightly bound, newly synthesized ATP to the ATP that has been demonstrated to be tightly bound to isolated α subunits of F_1 remains unclear, however; these are apparently not equivalent.[74–76]

These ideas constitute the basis for a new kind of conformational coupling hypothesis. The energy that is made available to the F_0F_1 system by proton efflux is viewed to drive a conformational change in F_1 that can bring about the release of the ATP that is readily synthesized but tightly bound on the surface of the enzyme.

Catalytic Cooperativity. Several independent lines of investigation have suggested that the mechanism of ATP synthesis and hydrolysis on the surface of F_1 may involve cooperative interactions between the subunits of this enzyme. These are not seen as being cooperative binding interactions, as are well known for allosteric proteins such as hemoglobin or regulatory enzymes. Instead, the sites are viewed as alternating between specific conformations, with catalytic events at one site promoting steps at the active site of a neighboring subunit. For example, chemical modification of only one tyrosine residue of F_1 causes complete inhibition of ATP hydrolysis,[77] suggesting ligand-induced asymmetry; when only one site is blocked, the others are somehow impaired. With bacterial F_1, the rate of dissociation of bound ADP has been found to be increased in the presence of ADP in solution,[78] a result that is interpretable in terms of nucleotide binding at one site promoting nucleotide release from another. A more sensitive probe of the effects of nucleotide binding on catalytic events is that of oxygen exchange measurements as carried out by Boyer and his colleagues.[73,79–81] These measurements of exchange of ^{18}O between water and P_i allow assessment of the number of dynamic reversals of ATP synthesis that occur on the surface of the enzyme, relative to the net catalytic turnover rate. The results indicate that the binding of ADP promotes the dissociation of ATP during net synthesis and that the binding of ATP promotes the release of ADP during net hydrolysis. Further compelling evidence for catalytic cooperativity is derived from the elegant experiments by Grubmeyer and Penefsky.[82] These workers found that occupancy of only one of the hydrolytic sites of F_1 by the radioactive "high-affinity" ATP analogue, 2′,3′-O-(2,4,6-trinitrophenyl)adenosine 5′-[γ-^{32}P]-triphosphate, is associated with a low rate of its hydrolysis. Addition of nonradioactive analogue

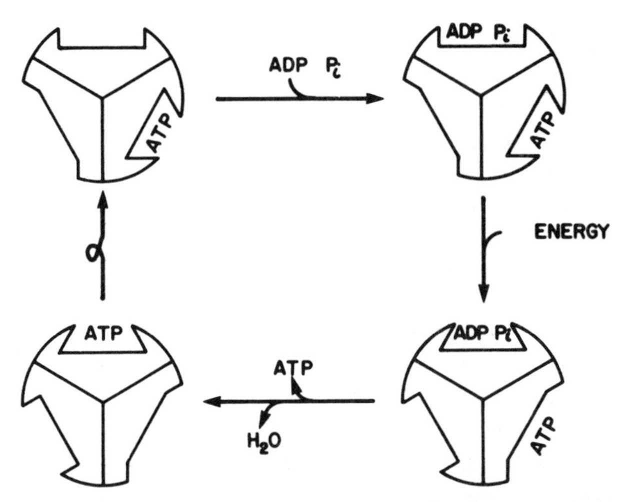

Figure 3.2. Model for ATP synthesis by F_1–ATPases that incorporates three alternating active sites per enzyme molecule, with catalytic cooperativity (see text). In this model, ATP synthesis per se is viewed as being reversible, but the product is so tightly bound that it cannot be released before the energy-driven reciprocal conformation change in the ATP binding sites. (Adapted from Reference 83.)

at higher concentrations (sufficient to occupy neighboring sites) results in a 15- to 20-fold increase in the rate of hydrolysis of the radioactive analogue.

A scheme incorporating all of the foregoing evidence and ideas, and assuming three catalytic sites per F_1 molecule, is presented in Figure 3.2.

REFERENCES

1. J. I. Harris and M. Waters, *The Enzymes,* vol. 13, P. D. Boyer, Ed., Academic, New York (1976), p. 1.
2. I. Krimsky and E. Racker, *Science* **122,** 319 (1955).
3. R. A. MacQuarrie and S. A. Bernhard, *J. Mol. Biol.* **55,** 181 (1971).
4. M. Buehner, G. C. Ford, D. Moras, K. W. Olsen, and M. G. Rossmann, *J. Mol. Biol.* **90,** 25 (1974).
5. D. Moras, K. W. Olsen, M. N. Sabeson, G. C. Ford, and M. G. Rossmann, *J. Biol. Chem.* **250,** 9137 (1975).
6. R. K. Scopes, *The Enzymes,* vol. 8, in P. D. Boyer, Ed., Academic, New York (1973), p. 335.
7. C. C. F. Blake, P. R. Evans, and R. K. Scopes, *Nature New Biol.* **235,** 195 (1972).

8. C. C. F. Blake and P. R. Evans, *J. Mol. Biol.* **84,** 585 (1974).
9. P. L. Wendell, T. N. Bryant, and H. C. Watson, *Nature New Biol.* **240,** 134 (1972).
10. T. N. Bryant, H. C. Watson, and P. L. Wendell, *Nature* **247,** 14 (1974).
11. R. D. Banks, C. C. F. Blake, P. R. Evans, R. Haser, D. W. Rice, G. W. Hardy, M. Merrett, and A. W. Phillips, *Nature* **279,** 773 (1979).
12. C. A. Pickover, D. B. McKay, D. M. Engelman, and T. A. Steitz, *J. Biol. Chem.* **254,** 11323 (1979).
13. C. T. Walsh and L. Spector, *J. Biol. Chem.* **246,** 1255 (1971).
14. A. Brevet, C. Roustan, G. Desvages, L.-A. Pradel, and N. van Thoai, *Eur. J. Biochem.* **39,** 141 (1973).
15. P. E. Johnson, S. J. Abbott, G. A. Orr, M. Sémériva, and J. R. Knowles, *Biochemistry* **15,** 2893 (1976).
16. J. R. Knowles, *Ann. Rev. Biochem.* **49,** 877 (1980).
17. M. R. Webb and D. R. Trentham, *J. Biol. Chem.* **255,** 1775 (1980).
18. B. D. Nageswara Rao, M. Cohn, and R. K. Scopes, *J. Biol. Chem.* **253,** 8056 (1978).
19. J. B. S. Haldane, *Enzymes,* Longmans, London (1930), Chapter V.
20. W. W. Cleland, *Biochim. Biophys. Acta* **67,** 104 (1963).
21. B. D. Nageswara Rao, D. H. Buttlaire, and M. Cohn, *J. Biol. Chem.* **251,** 6981 (1976).
22. J. L. Robinson and I. A. Rose, *J. Biol. Chem.* **247,** 1096 (1972).
23. B. D. Nageswara Rao, F. J. Kayne, and M. Cohn, *J. Biol. Chem.* **254,** 2689 (1979).
24. R. K. Gupta and A. S. Mildvan, *J. Biol. Chem.* **252,** 5967 (1977).
25. D. Dunaway-Mariano, J. L. Benovic, W. W. Cleland, R. K. Gupta, and A. S. Mildvan, *Biochemistry* **18,** 4347 (1979).
26. D. I. Stuart, M. Levine, H. Muirhead, and D. K. Stammers, *J. Mol. Biol.* **134,** 109 (1979).
27. J. L. Wyatt and R. F. Colman, *Biochemistry* **16,** 1333 (1977).
28. A. E. Annamalai and R. F. Colman, *J. Biol. Chem.* **256,** 10276 (1981).
29. G. A. Orr, J. Simon, S. R. Jones, G. J. Chin, and J. R. Knowles, *Proc. Natl. Acad. Sci. (USA)* **75,** 2230 (1978).
30. W. A. Blätter and J. R. Knowles, *Biochemistry* **18,** 3927 (1979).
31. W. A. Bridger, in *The Enzymes,* vol. X, P. D. Boyer, Ed., Academic, New York (1974), Chapter 18.
32. W. A. Bridger, W. A. Millen, and P. D. Boyer, *Biochemistry* **7,** 3608 (1968).
33. P. D. Boyer, *Science* **141,** 1147 (1963).
34. R. A. Mitchell, L. G. Butler, and P. D. Boyer, *Biochem. Biophys. Res. Commun.* **16,** 545 (1964).
35. J. S. Nishimura and A. Meister, *Biochemistry* **4,** 1457 (1965).

36. M. Cohn, *J. Cell. Comp. Physiol.* **54,** 157 (1959).
37. W. A. Bridger, *Biochem. Biophys. Res. Commun.* **42,** 948 (1971).
38. E. R. Brownie and W. A. Bridger, *Can. J. Biochem.* **50,** 719 (1972).
39. R. F. Ramaley, W. A. Bridger, R. W. Moyer, and P. D. Boyer, *J. Biol. Chem.* **242,** 4287 (1967).
40. F. J. Moffet, T. Wang, and W. A. Bridger, *J. Biol. Chem.* **247,** 8139 (1972).
41. G. S. Bild, C. A. Janson, and P. D. Boyer, *J. Biol. Chem.* **255,** 8109 (1980).
42. W. T. Wolodko, M. D. O'Connor, and W. A. Bridger, *Proc. Natl. Acad. Sci. (USA)* **78,** 2140 (1981).
43. H. J. Vogel and W. A. Bridger, *J. Biol. Chem.* **257,** 4834 (1982).
44. W. A. Bridger, *Can. J. Biochem.* **59,** 1 (1981).
45. P. D. Boyer, B. Chance, L. Ernster, P. Mitchell, E. Racker, and E. C. Slater, *Ann. Rev. Biochem.* **46,** 955 (1977).
46. P. Mitchell, *Chemiosmotic Coupling in Oxidative and Photosynthetic Phosphorylation,* Glynn Research, Ltd., Bodmin, U.K. (1966).
47. T. E. Conover and G. F. Azzone, in *Mitochondria and Microsomes,* C. P. Lee, G. Schatz, and G. Dallner, Eds., Addison-Wesley, Reading, MA (1981), p. 481.
48. L. Ernster, R. W. Estabrook, and E. C. Slater, Eds., *Dynamics of Energy-Transducing Membranes,* Elsevier, Amsterdam (1974).
49. E. Racker, *A New Look at Mechanisms in Bioenergetics,* Academic, New York (1976).
50. P. Mitchell, *Science* **206,** 1148 (1979).
51. R. Fillingame, *Ann. Rev. Biochem.* **49,** 1079 (1980).
52. A. E. Senior, in *Membrane Proteins in Energy Transduction,* R. A. Capaldi, Ed., Marcel Dekker, New York (1979), p. 233.
53. S. D. Dunn and L. A. Heppel, *Arch. Biochem. Biophys.* **210,** 421 (1981).
54. J. C. Brooks and A. E. Senior, *Arch. Biochem. Biophys.* **147,** 467 (1976).
55. R. J. VandeStadt, B. L. DeBoer, and K. VanDam, *Biochim. Biophys. Acta* **292,** 338 (1973).
56. P. D. Bragg, P. L. Davies, and C. Hou, *Arch. Biochem. Biophys.* **159,** 664 (1973).
57. N. Nelson, B. I. Kanner, and D. L. Gutnick, *Proc. Natl. Acad. Sci. (USA)* **71,** 2720 (1974).
58. M. Futai, P. C. Sternweis, and L. A. Heppel, *Proc. Natl. Acad. Sci. (USA)* **71,** 2725 (1974).
59. G. Vogel and R. Steinhart, *Biochemistry* **15,** 208 (1976).
60. P. C. Sternweis, *J. Biol. Chem.* **253,** 3123 (1978).
61. M. Yoshida, H. Okamoto, N. Sone, H. Hirata, and Y. Kagawa, *Proc. Natl. Acad. Sci. (USA)* **74,** 936 (1977).
62. P. C. Steinweis and J. B. Smith, *Biochemistry* **19,** 526 (1980).
63. S. D. Dunn and M. Futai, *J. Biol. Chem.* **255,** 113 (1980).

64. S. D. Dunn, *J. Biol. Chem.* **255,** 11857 (1980).
65. H. H. Paradies, *FEBS Lett.* **120,** 289 (1980).
66. P. D. Bragg and C. Hou, *Arch. Biochem. Biophys.* **178,** 486 (1977).
67. P. D. Bragg and C. Hou, *Biochem. Biophys. Res. Commun.* **72,** 1042 (1976).
68. J. V. Moroney and R. E. McCarty, *J. Biol. Chem.* **254,** 8951 (1979).
69. E. C. Slater, *Nature* **172,** 975 (1953).
70. M. Webb, C. Grubmeyer, H. S. Penefsky, and D. R. Trentham, *J. Biol. Chem.* **255,** 11637 (1980).
71. P. D. Boyer, in *Proc. Int. Symp. Enzyme Chem., Tokyo & Kyoto,* Maruzen, London (1957), p. 301.
72. J. A. Russo, C. M. Lamos, and R. A. Mitchell, *Biochemistry* **17,** 473 (1978).
73. C. Kayalar, J. Rosing, and P. D. Boyer, *J. Biol. Chem.* **252,** 2486 (1977).
74. P. D. Boyer, M. Gresser, C. Vinkler, D. Hackney, and G. Choate, in *Structure and Function of Energy-Transducing Membranes,* K. van Dam and B. F. van Gelder, Eds., Elsevier, Amsterdam (1977), p. 261.
75. M. Gresser, J. Cardon, G. Rosen, and P. D. Boyer, *J. Biol. Chem.* **254,** 10649 (1979).
76. R. E. McCarty, *Trends Biochem. Sci.* **4,** 28 (1979).
77. S. Ferguson, W. J. Lloyd, and G. K. Radda, *Eur. J. Biochem.* **54,** 127 (1975).
78. R. Adolfsen and E. N. Moudrianakis, *Arch. Biochem. Biophys.* **172,** 425 (1976).
79. D. D. Hackney and P. D. Boyer, *J. Biol. Chem.* **253,** 3164 (1978).
80. G. L. Choate, R. L. Hutton, and P. D. Boyer, *J. Biol. Chem.* **254,** 286 (1979).
81. R. L. Hutton and P. D. Boyer, *J. Biol. Chem.* **254,** 9990 (1979).
82. C. Grubmeyer and H. S. Penefsky, *J. Biol. Chem.* **256,** 3728 (1981).
83. R. L. Cross, *Ann. Rev. Biochem.* **50,** 681–714 (1981).

4

Mechanisms of ATP Utilization

Adenosine triphosphate is consumed by a huge variety of biological processes, and it would be a tall order indeed to discuss the details of ATP utilization in anything approaching a comprehensive manner. In this chapter we consider a few representative examples of different kinds of systems that utilize ATP. Those to be discussed include Na^+–K^+–ATPase as a representative of the transport ATPases, myosin ATPase to illustrate the means of utilization of ATP chemical bond energy for mechanical work, and hexokinase as a good example of the thoroughly investigated and relatively well understood enzymes of intermediary metabolism that consume ATP.

SODIUM–POTASSIUM ATPase

This enzyme is responsible for the transport of Na^+ from cells and the concomitant transport of K^+ from the outside to the inside. Since both of these transport processes are against a concentration gradient, energy is required, and this is provided by the hydrolysis of ATP. Active transport of Na^+ and K^+ consumes a very significant fraction of the ATP that is produced by most animal cells. It is fundamental for the maintenance of cell volume, the excitability of nerves, and absorption processes in specialized tissues such as kidney and intestine. In the kidney which is responsible for excretion of excess Na^+ into the urine while retaining K^+ in the filtered blood, over one-half of the ATP that is generated by catabolic processes is consumed by the action of renal Na^+–K^+–ATPase.

As would be expected, the Na^+–K^+–ATPase molecule is a component of the plasma membrane of animal cells. The enzyme has been studied, usually following solubilization by means of detergent, from a variety of sources. These include, in particular, Na^+–K^+–ATPase from the renal medulla and brain, tissues that are rich in this activity because of their physiological function. Also, students of Na^+–K^+–ATPase have taken advantage of the exceptionally high levels of enzyme that are found in specialized organs such as the electric organ of the eel, *Electophorus electicus*. It is interesting that the properties of the enzymes from diverse sources are very similar, suggesting that the enzyme developed in the earliest stages of evolution of multicellular organisms and that very little modification of its structure is required to satisfy the particular needs of a given tissue.

Specificity

The enzyme Na^+–K^+–ATPase catalyzes the hydrolysis of ATP to ADP and P_i, but only in the combined presence of three cations, namely, Mg^{2+}, Na^+, and K^+.[1] The activity of the enzyme is inhibited by the so-called cardiac glycoside, ouabain; this is useful in many ways for study of the Na^+–K^+–ATPase. For example, one can separate Na^+–K^+–ATPase activity from other contaminating ATPases by virtue of its sensitivity to ouabain. Thus ouabain-sensitive hydrolysis of CTP[2–4] and ITP[3,4] have been demonstrated, for example, establishing that the enzyme does not have strict specificity for ATP. The relative rates of hydrolysis of various nucleoside triphosphates have been reported[5] to be 100:49:2.3:2.4:$<$0.6:$<$0.6:$<$0.6 for ATP, dATP, CTP, ITP, GTP, UTP, and TTP, respectively. Both Na^+ and K^+ are required for activity; the Na^+ requirement seems to be absolute, but various monovalent cations can satisfy the K^+ requirement. In order of effectiveness, these are $Tl^+ > K^+ > Rb^+ > NH_4^+ > Cs^+ > Li^+$.[6–8] As indicated above, Mg^{2+} is also required for activity, as it is for virtually all enzymes that use ATP. The requirement for Mg^{2+} can also be satisfied by other divalent cations such as Mn^{2+} and Co^{2+}.[9–11]

Mechanism of ATP Hydrolysis

In contrast to F_1–ATPase (see Chapter 4) and myosin ATPase (see below), the mechanism of action of Na^+–K^+–ATPase has been shown to involve the intermediate participation of phosphorylated enzyme intermediates. One challenge for students of this enzyme is to correlate the steps in the reaction to the physiological function, namely, the active transport of monovalent cations through the membrane. The postulated steps of the reaction are as follows:[1]

$$E_1 + ATP \underset{}{\overset{Mg^{2+}\cdot Na^+}{\rightleftharpoons}} ADP + E_1\text{–}PO_3 \tag{2}$$

$$E_1\text{–}PO_3 \overset{Mg^{2+}}{\rightleftharpoons} E_2\text{–}PO_3 \tag{3}$$

$$E_2\text{–}PO_3 + H_2O \overset{K^+}{\rightleftharpoons} E_2 + P_i \tag{4}$$

$$E_2 \rightleftharpoons E_1 \tag{5}$$

Here, E_1 and E_2 designate different enzyme conformations; it has been suggested that E_2 may have a higher affinity for K^+.[12]

It has been shown[5,13,14] that ATP can bind to the free enzyme in the absence of Mg^{2+} and that such binding can be prevented by preincubation of the enzyme with either ouabain or K^+; Na^+ antagonizes this effect of K^+ without having a direct effect alone. Studies with various analogues of ATP suggest that the 6-amino group and the γ phosphoryl of the nucleotide are particularly crucial for its recognition by the enzyme.[5]

The phosphorylated derivative of Na^+–K^+–ATPase has received a great deal of attention over the last two decades. The first indication of a covalent E–PO_3 form of the enzyme was based on catalysis of an ATP–ADP exchange by a partially purified nerve Na^+–K^+–ATPase preparation.[15] This was borne out by the widespread demonstrations that Na^+–K^+–ATPase from various sources could incorporate precipitable ^{32}P from [$\gamma^{32}P$]ATP in the presence of Mg^{2+} and K^+.[16–23] The presence of K^+ during this treatment was found to greatly reduce the extent of phosphorylation, and the subsequent addition of K^+ resulted in a rapid hydrolysis of radioactive P_i from the enzyme, in keeping with equation (4) above. The presence of ouabain was found to inhibit both the phosphorylation and dephosphorylation steps.[24] These and other experiments[1] strongly suggested that the phosphoenzyme may be a catalytic intermediate in the course of ATP hydrolysis, and the chemical nature of the phosphoenzyme form has thus received considerable attention. Acyl phosphates can be distinguished as a class by their lability at both extremes of pH and by their sensitivity to neutral hydroxylamine treatment. The phosphorylated intermediate of Na^+–K^+–ATPase shows these properties,[21] suggesting the presence of either glutamyl or aspartyl phosphate in the enzyme. The first attempt to distinguish between these two possibilities was based on the use of a radioactive derivative of hydroxylamine followed by chromatographic resolution of the pronase-digested enzyme,[25] thus suggesting that the site of phosphorylation was a glutamyl residue. A more direct method, however, first used to identify β-aspartyl phosphate at the active site of Ca^{2+}-activated ATPase of sarcoplasmic reticulum,[26] involves reductive cleavage of the phosphorylated enzyme with borohydride; β-aspartyl phosphate is converted to homoserine by this treatment, whereas γ-glutamyl phosphate would give rise to α-amino-δ-hydroxyvaleric acid. Homoserine was identified[27] in the digests of the borohydride-reduced phosphorylated Na^+–K^+–ATPase,[27] clearly indicating that the site of phosphorylation is an aspartyl residue.*

The phosphorylated form of Na^+–K^+–ATPase has properties that are expected for a true catalytic intermediate. For example, the rate of turnover and the pre-steady-state rate of formation of the phosphoenzyme are both consistent with its participation in the overall reaction.[28,29] Furthermore, Fossel et al.[30] have recently studied the phosphoenzyme form of Na^+–K^+–ATPase from the salt gland of the duck by means of high-resolution ^{31}P-NMR. Because of the insensitivity of this technique and the unavoidable difficulties in dealing with an enzyme of membrane origin, the resonance that these workers have attributed to the β-aspartyl phosphate residue is weak and at +17.4 ppm almost overlaps with the resonance of the β-phosphoryl group of ATP.† Therefore, detection of the resonance was facilitated by the fact that the enzyme can be phosphorylated not only in the presence of ATP, but also by incubation with P_i and Mg^{2+} in the presence of ouabain.[31] As expected for the catalytic role assigned to the E–PO_3 form, the resonance disappears when K^+ is added to the sample, indicating that K^+ stimulates the hydrolysis of the acyl–phosphate linkage at the

*The earlier erroneous identification of γ-glutamyl phosphate[25] may be attributable to incomplete digestion of the treated protein with pronase.

†This chemical shift for the β-aspartyl phosphate residue is outside the range expected from examination of model compounds (2 ppm for acetyl phosphate and +11.6 ppm for β-phosphoaspartic acid). Phosphorylation of the dipeptide seryl aspartate gave rise to a resonance at 17.4 ppm,[30] emphasizing the notable sensitivity of the ^{31}P chemical shifts of phosphate anhydride compounds to their nearby chemical environment.

active site. K^+-dependent hydrolysis of the phosphoenzyme accounts in part for the K^+-dependent, ouabain-inhibitable nonspecific phosphatase activity that is demonstrated by Na^+–K^+–ATPase from a variety of sources. (For further discussion, see Dahl and Hokin.[1])

The transport of Na^+ and K^+ that is catalyzed by Na^+–K^+–ATPase has an unusual stoichiometry. For red-cell,[32-34] nerve,[35] and renal tubule[36] Na^+–K^+–ATPases, the ratio of Na^+ ions transported outside to K^+ ions transported inside to the number of ATP molecules hydrolyzed has been found to be 3:2:1.

Subunit Structure and Membrane Orientation

The enzyme Na^+–K^+–ATPase is composed of two different subunits. The larger of the two subunits has a molecular weight in the range 95,000–100,000 and contains the site that is phosphorylated during the course of catalysis.[37-40] Furthermore, experiments in which various substrates and effectors were tested for their ability to influence the reaction of the enzyme with N-ethyl maleimide[41] strongly suggest that the larger subunit contains not only the site for ATP hydrolysis, but also sites for interaction with Na^+, K^+, and ouabain. Effects on sulfhydryl reactivity indicate that the conformation of this polypeptide is markedly affected by phosphorylation.[41] The presence of binding sites for ATP and Na^+ clearly implies that this subunit has access to the cytoplasmic side of the plasma membrane, a concept borne out by the fact that antibodies to the large subunit bind specifically to the inner surface of this membrane.[42] On the other hand, its sites for ouabain and K^+ suggest that this polypeptide can also interact with ligands on the outside of the cell. It is reasonable to conclude that the larger subunit spans the membrane, perhaps with conformational changes allowing sequential access of its cation binding sites to the inner and outer milieu.

The smaller subunit is a glycoprotein with a molecular weight of about 55,000.[38] The presence of conjugated sialic acid residues suggests that this subunit forms a component of the cell surface. Active Na^+–K^+–ATPase reconstituted into phosphatidylcholine vesicles contains equimolar amounts of the two subunits,[36] confirming that both are required for assembly of a functional cation transporting system.

With this picture of sites, conformational changes, subunit structure, and membrane location, it is easy to visualize models for transport of Na^+ and K^+. In most of these[1,43] the enzyme is postulated to exist in two alternating conformations. One of these represents the dephosphorylated enzyme with the high-affinity Na^+ sites facing the cytoplasm. When these are occupied by Na^+ ions, ATP-dependent phosphorylation and conformational change can proceed. As a result of this shape change, the Na^+ ions are transported to the outer surface and the tenacity of their binding is reduced, and these Na^+ ions then dissociate to be replaced (not necessarily at exactly the same sites) by K^+ ions. Attachment of K^+ is accompanied by the well-known stimulation of the dephosphorylation of the enzyme, thus converting it back to the original conformation, with the cation-binding sites on the inside and with low affinity for K^+ and high affinity for Na^+. The net result of this full cycle is the ATP-dependent pumping of Na^+ and K^+ in opposite directions.

Further discussion of K^+ transport equilibria in muscle, nerve, mitochondrial, and epithelial cell membranes is presented in an earlier volume in this series.[44]

ATP HYDROLYSIS BY MYOSIN AND ACTOMYOSIN

One of the most challenging problems in modern biochemistry is the elucidation of the mechanisms of ATP utilization for mechanical work and the regulation of this process by effectors such as Ca^{2+} ions. In addition to contraction of muscle, ATP-dependent contractile systems are at work in a variety of other processes, including chromosomal spindle formation, cytoplasmic streaming, and movement of cilia and flagella. In all these cases, a sliding filament model involving ATP-dependent translocation of actin and myosin molecules has been implicated as the basis for contraction. Detailed and comprehensive reviews of these systems have recently been published.[45,46]

The thick filaments of muscle are assemblies of myosin molecules. The globular head region of the myosin molecule may be prepared either by limited trypsin treatment with the production of heavy meromyosin (HMM) or by following this by digestion with pepsin, releasing intact globular heads (subfraction 2 or *S*2). The *S*2 fragments contain the site responsible for hydrolysis of ATP, together with the point of interaction between actin and myosin that is involved in generating the power stroke of muscle contraction (see below).

In resting muscle under physiological ionic conditions, hydrolysis of ATP is very slow, with a first-order rate constant for hydrolysis of 0.025 sec^{-1}. When myosin is tested *in vitro* under comparable conditions, there is a rapid early burst of ATP hydrolysis, followed by the much lower steady-state rate that is comparable to that observed in resting fibers. During this burst phase, roughly one P_i molecule is released per myosin active site.[47,48] One explanation for the stoichiometric burst of product formation could be the presence of a covalent intermediate. Attempts to detect the presence of a phosphorylated intermediate have not been fruitful,[45] and the absence of a covalent intermediate has been confirmed by observations on the stereochemical course of phosphoryl transfer catalyzed by myosin ATPase.[49] When ATP-γS that is stereospecifically labeled in the γ position with ^{18}O is then hydrolyzed in the presence of ^{17}O-enriched water, the chirality of the product indicates that the hydrolysis proceeds with stereochemical inversion, consistent with a direct in-line attack of water on the terminal phosphoryl group of ATP.

If there is no covalent intermediate, what is the explanation for the rapid release of product in the pre-steady-state period of catalysis by myosin ATPase? Further clues were provided by ^{18}O exchange measurements performed by Bagshaw and Trentham,[50] who found that the rate of exchange of oxygen between water and P_i is much more rapid than net P_i production by myosin ATPase. An interpretation of this result is that the reaction

$$ATP + H_2O \rightleftharpoons ADP + P_i$$

is readily reversible, with many resyntheses of ATP (without spatial selectivity of phosphate oxygens) occurring per net reaction. This concept was confirmed by the elegant experiments of Geeves et al.,[51] who showed that myosin catalyzes slow scrambling of oxygen atoms between the β,γ-bridge and the β-nonbridge positions of ATP. For this scrambling to occur, the bond between the β- and γ-phosphoryl groups must be hydrolyzed and then resynthesized with a ramdom choice of any of the three β-phosphoryl oxygens of ADP to attack P_i and thus form the bridge atom. Not only are these oxygens scrambled; Sleep et al.[52] have shown that the four oxygens of the bound product P_i have an equal probability of being displaced by the attacking ADP in the resynthesis of ATP at the *S*1-active site. Tumbling or rotation of the still-bound P_i thus occurs more rapidly than the reversal of the hydrolysis step. Taking all of the oxygen exchange data together (i.e., both the positional scrambling and the P_i:H_2O exchange), it seems clear that the reaction catalyzed by myosin ATPase is readily reversible on the enzyme surface, and the fact that ADP and P_i can be reconverted to ATP before they dissociate from the enzyme surface has important implications for the mechanism of ATP-dependent interaction between actin and myosin (see below).

If the rate of ATP resynthesis exceeds the rate of dissociation of ADP and P_i, the rate-limiting step must be the release of products or some other intermediate slow step that follows the hydrolysis reaction. Lymn and Taylor[48] proposed that product release is rate limiting and that the activating effect of actin on myosin ATPase could be related to a stimulation of product release. This is supported by the observations[48] that the intermediate myosin·ADP complex can be separated from reactants in solution by rapid fractionation techniques and that the rate of decay of this complex is equivalent to the overall steady-state rate of ATP hydrolysis. This is likely an oversimplification, however, in view of the fact that direct measurements by equilibrium dialysis or by the use of chromophoric ATP analogues have led to a measurement of the rate of dissociation of ADP that greatly exceeds the steady-state rate of the reaction.[53,54] These and a variety of complex kinetic considerations and physical measurements[45] have indicated that the simplest scheme for representation of the steps of ATP hydrolysis must require at least seven steps involving three conformations for myosin:

$$\mathrm{M + ATP \rightarrow M{\cdot}ATP \rightarrow M{*}{\cdot}ATP \rightarrow M{**}{\cdot}ADP{\cdot}P_i \rightarrow}$$
$$\mathrm{M{*}{\cdot}ADP{\cdot}P_i \rightarrow M{*}{\cdot}ADP + P_i \rightarrow M{*}{\cdot}ADP \rightarrow M + ADP}$$

Here, the conversion of $\mathrm{M{**}{\cdot}ADP{\cdot}P_i}$ to $\mathrm{M{*}{\cdot}ADP{\cdot}P_i}$ is viewed as the rate-limiting step, accounting for the reversal of ATP hydrolysis and for the fact that ADP that is bound from solution (rather than that produced from ATP cleavage) dissociates relatively rapidly.

The foregoing discussion is based on measurements with myosin subfragment 1 (*S*1), but they provided a framework for ideas and further experimentation that have led to proposals for the coupling of ATP hydrolysis to mechanical work. Since the power stroke of muscle contraction is followed by actin dissociation from the myosin heads and subsequent reattachment, the problem is to determine at which points the actin association and dissociation intervene in the overall scheme for ATP hydrolysis. Many proposals have been advanced, but current views generally support the model proposed by Lymn and Taylor,[55] with the inevitable refinements. In essence, all schemes for the sliding filament involve four steps, namely, the dissociation of the myosin head from the thin filament complex, movement of the myosin head by conformational change, recombination of the head with actin, and the power stroke. By the Lymn–Taylor model,[55] these are coupled to ATP hydrolysis as follows. The binding of ATP is accompanied by very rapid dissociation of actin ($k < 1000\ \mathrm{sec^{-1}}$). The hydrolysis of ATP then occurs on the *dissociated* myosin head in stages that include a fast conformational change (equivalent to $\mathrm{M{\cdot}ATP \rightarrow M{*}{\cdot}ATP}$) with a rate constant of 400 $\mathrm{sec^{-1}}$, and the actual hydrolysis step ($k = 100$–$150\ \mathrm{sec^{-1}}$). Actin then combines with the long-lived myosin-product complex ($M^{**}\cdot ADP\cdot P_i$) with a second-order rate constant of $3 \times 10^5\ M^{-1}/sec$, and then ADP and P_i can dissociate rapidly ($k = 10\text{–}20\ sec^{-1}$). The cycle is then completed by a second round of ATP binding and actin dissociation.

A key feature of the basic model advanced by Lymn and Taylor, or any refinement of it, is that ATP hydrolysis occurs on the myosin head when it is not in contact with the thin filament. If ADP and P_i dissociated readily from the site at which this cleavage takes place, the system would become a simple ATPase with no potential for mechanical work since the actual hydrolysis step is isolated in time from the power stroke. It is intriguing, therefore, that two processes that are so fundamental as muscle contraction and oxidative phosphorylation (see Chapter 3) have evolved to make use of the same principle: energy can be captured in protein conformations that bind nucleotides very tightly. In the case of muscle contraction, ATP is hydrolyzed reversibly, but the product ADP and P_i can dissociate only after a highly exergonic conformational change occurs that is coupled to formation of the actin–myosin cross bridge. In the somewhat reciprocal case of oxidative phosphorylation, ATP is thought to be formed reversibly on the surface of the enzyme, but it cannot be released without a conformational change that is driven by the exergonic proton flux. The broad similarities in the principles that are basic to these processes are striking.

HEXOKINASE

One of the most extensively studied phosphotransferases is yeast hexokinase. It catalyzes the phosphorylation of glucose as follows:

$$\text{Glucose} + \text{ATP} \rightleftharpoons \text{Glucose-6-phosphate} + \text{ADP}$$

There has been considerable animated controversy over the kinetic mechanism of yeast hexokinase regarding whether it is sequential or ping pong and ordered or random. Early steady-state kinetic studies,[56] performed just at the time of development of modern enzyme kinetics, indicate that the mechanism is sequential, meaning that both substrates must attach to their respective sites before phosphoryl transfer occurs. This is to be distinguished from ping-pong kinetics, in which a double displacement reaction occurs, with participation of a free phosphorylated enzyme intermediate. The evidence in favor of sequential kinetics is the appearance of an intersecting pattern of reciprocal plots (v^{-1} versus S^{-1})

at various fixed concentrations of the second substrate. Although there is wide agreement that the hexokinase reaction follows a sequential pathway, there has been an inordinate amount of controversy over the order of substrate addition within the sequential framework. Either a random or ordered addition of substrates to the enzyme could produce the intersecting pattern observed; techniques such as product inhibition and alternative substrate inhibition must be used to distinguish between these two possibilities. For example, studies of the effects of mannose and fructose on the rate of glucose phosphorylation were interpreted in terms of an ordered sequence with glucose binding before ATP,[57] and the inhibition patterns of an ATP analogue with both glucose and ATP were similarly taken as evidence for an ordered sequence with only glucose combining with the free enzyme.[58] On the other hand, the kinetics of isotope exchange at equilibrium are not compatible with a strictly ordered sequence.[59] The rates of glucose:glucose-6-phosphate exchange and ATP:ADP exchange both show hyperbolic increases as either substrate–product pair is elevated in concentration. (If the mechanism were ordered with glucose on first and glucose-6-phosphate off last, the equilibrium exchange between these two reactants should be supressed at saturating [ATP] and [ADP].) This has been explored further by Britten and Clark,[60] who measured the ratio of the two exchange rates as a function of increasing [glucose] and [ATP] at equilibrium. They found that glucose has no effect on the ratio but that there is a fivefold effect of ATP. These are the effects expected for a *preferred* pathway, formally random but with glucose the first substrate to add at least 90 percent of the time:

E · G   E · G6P

E   E · G · ATP ⇌ E · G6P · ADP   E

E · ATP   E · ADP

The foregoing discussion does not shed much light on the more fundamental aspects of catalysis by this enzyme, namely, the mechanism of phosphoryl transfer. Although the sequential kinetics may be interpreted straightforwardly as an indication that there is no covalent phosphoenzyme intermediate, all that can be concluded on the basis of this kind of information is that there is no *free* phosphoenzyme that has a significant lifetime without bound substrates. In fact, it has been clearly demonstrated that an enzyme that is known to have a phosphoenzyme intermediate, succinyl–CoA synthetase, still exhibits sequential kinetics.[61] This is because of a phenomenon known as *substrate synergism*[62] that describes the property of an enzyme that must have all substrate-binding sites occupied before it becomes an effective catalyst for a partial reaction. So one can still postulate the possible existence of a form of hexokinase in the central complex with transient formation of a phosphoenzyme as a component of a double-displacement mechanism. In fact, certain observations have been taken as evidence in favor of a phosphoenzyme. These include the intrinsic ATPase activity (40,000 times slower than glucose phosphorylation),[63] but the fact that this hydrolytic activity is stimulated by the nonsubstrate sugar xylose suggests that the ATPase is simply an "accident" that results when water is the only potential phosphate acceptor present at the active site. Other observations include a very slow ATP:ADP exchange in the absence of added hexose (which could always be the result of a contaminant) and actual identification of a phosphoenzyme when hexokinase is incubated with labeled ATP and *D*-xylose.[64] Although this phosphoenzyme is catalytically inactive (i.e., it does not turn over during subsequent catalysis), one could invoke various activation phenomena to press the data into a mold that is consistent with a phosphoenzyme intermediate. In reviewing this literature, Jeremy Knowles[64a] has observed:

> Nearly every phosphokinase that has been scrutinized has its complement of slow partial reactions; the problem is to decide whether they are misleading artefacts or useful clues pointing to a phosphoenzyme.

In the case of hexokinase, these observations now appear to have been misleading. There is now a wealth of harder data that point clearly to the absence of a phosphoenzyme. One novel approach is that of positional isotope scrambling, as devised by Midelfort and Rose.[65] These workers reasoned that a possible reason for failure to detect a phosphoenzyme intermediate might be that the coproduct ADP is bound so tightly that it cannot dissociate sufficiently for detectable ATP:ADP exchange. If reversible phosphorylation by ATP does occur even without ADP dissociation, however, this should result in a scrambling of the β,γ-bridge oxygen of ADP into the β-nonbridge position. Rather heroic experimental design and execution demonstrated no such scrambling, arguing very strongly against even fleeting phosphorylation of hexokinase by ATP.* But the strongest evidence against the existence of a phosphorylated intermediate is again derived from stereochemical experiments; the transfer of phosphoryl from ATP to glucose proceeds with inversion of stereochemical configuration at phosphorus,[66-68] indicating that the transfer occurs directly from donor to acceptor

*It could still be argued that the phosphorylation is *so* transient that the β-phosphoryl group of ADP does not have time to rotate during its lifetime.

within the central complex without intermediate covalent involvement of the enzyme. As can be seen when we turn our attention to the crystal structure of hexokinase, the absence of a phosphoenzyme intermediate has implications for conformational transitions in the catalytic mechanism.

Yet another ingenious probe for exploring the mechanism of action of hexokinase has been devised by Rose and his colleagues. Wilkinson and Rose[69] have applied a clever rapid quench and isotope trapping technique to detect the levels of intermediate species during catalysis. When hexokinase was rapidly mixed with labeled glucose and ATP and the reaction was quenched with acid after a few milliseconds, a burst of glucose-6-phosphate (G6P) was produced that is equivalent to about one-half the molar concentration of enzyme. This indicates that about half of the enzyme molecules have bound reactants that have already undergone transfer of phosphoryl from ATP to glucose. If the quench is delayed for some time, the subsequent rate of net production of G6P is equivalent to the steady-state velocity. Together with other experiments that establish that the concentrations of binary enzyme·sugar complexes are very low (i.e., that the rate of G6P release is limited by the rate of dissociation of ADP), these observations show that during the steady state about half of the enzyme exists as the E·G6P·ADP complex. Therefore, the equilibrium constant for interconversion of E·G·ATP and E·G6P·ADP must be near unity. It is striking that similar quantitative conclusions have been reached with other enzymes such as phosphoglycerate kinase and pyruvate kinase and that rapid equilibration of central complexes has been indicated for F_1 and myosin ATPases (see Chapter 3). Clearly, one kind of strategy for effective catalysis that has been adopted by at least a broad variety of phosphotransferases is to compensate for the overall equilibrium constant by adjustments in the tightness of binding of substrates.

There are some important lessons to be learned from a review of the crystal structures of hexokinase and its complexes with substrates, as elucidated by Steitz and his colleagues. Yeast hexokinase is a dimer of chemically identical subunits of 50,000 molecular weight.[70] Early crystallographic studies at relatively low resolution[71] led to an interesting and unexpected finding: although the two subunits have the same amino acid sequence and virtually equivalent secondary and tertiary structures, they are assembled into an asymmetrical dimeric structure, with the subunit contacts shifted over 10 Å from what would have been the twofold molecular axis. A single molecule of ATP, apparently far removed from the active site and not partipating at all in catalysis, is located at the intersubunit contact. The subunits themselves are folded into two principal domains, the large lobe (residues 1–57 and 192–458) and the small lobe (residues 58–191). The active site is located in a cleft between these two domains.

Improved resolution and refinement of the structures has recently allowed more detailed scrutiny of the structural transitions that accompany binding of substrates.[72–74] The binding site for ATP at the active site has been explored by examination of a difference electron density map of hexokinase crystals that had been soaked in the analogue 8-bromoadenylate.[74] The map showed the ribose to be anti relative to the adenine, which is itself bound in a shallow depression on the surface of the large lobe near the cleft between the two domains. A model for this derivative of ATP bound at the active site has been constructed from the foregoing structure combined with the crystal structure of tripolyphosphato tetraamine cobalt(III), and this is indicated in Figure 4.1. This structure shows the extended triphosphate chain of ATP stretching away from the adenosine

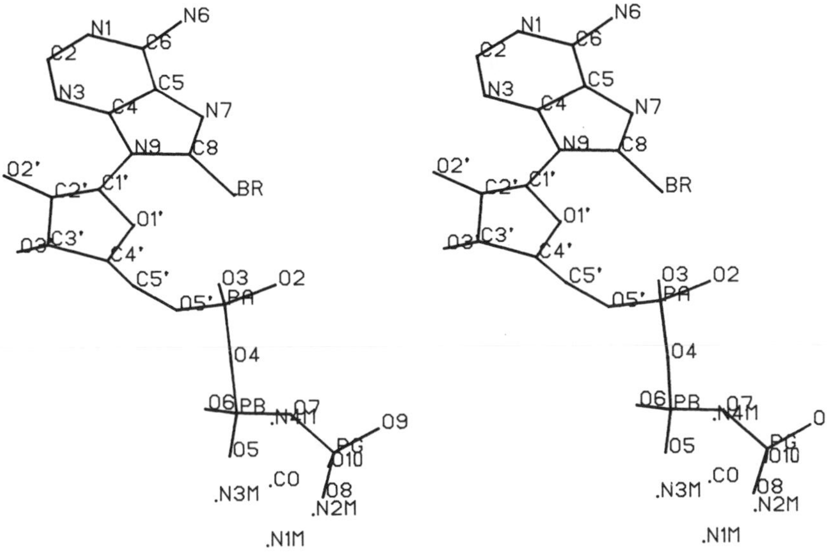

Figure 4.1. Stereo model for the cobalt complex of 8-bromo-ATP bound to the active site of yeast hexokinase. The model was constructed with the aid of a MMS-X vector graphics molecular modeling system at the University of Alberta,[76] using the coordinates as tabulated by Shoham and Steitz.[74] Identification of atoms is obvious, with the exceptions of PA, PB, and PC, which represent the α-, β, and γ-phophoryl groups of the nucleotide, and N1M–N4M, which represent the positions of the tetraamine substituents on the cobalt atom. The phosphoryl acceptor (the 6-hydroxyl group of glucose) is 6 Å away from the γ-phosphoryl group in the crystal structure, too far for direct attack (see text for further details).

portion of the molecule toward the binding site for glucose. It is of great interest, however, that the oxygen-6 of glucose in the crystal structure is beyond reach of the γ-phosphoryl group of ATP, some 6 Å away but in line. Since this distance is too great for direct nucleophilic attack of the phosphoryl group by the glucose hydroxyl oxygen, two explanations are possible. One would involve the possible intermediate participation of a phosphorylated amino acid side chain in a double-displacement mechanism, but that possibility seems to be excluded by the experiments described above, especially by the observed inversion of stereochemical configuration that accompanies the overall reaction. Moreover, there is no obvious electron density in the 6-Å gap that one could assign to a putative phosphoryl transferring moiety. The second and favored explanation is that hexokinase undergoes a major conformational change when glucose is bound at the active site.[73,75] This structural transition involves a rotation of the small lobe by 12° relative to the large lobe, moving the polypeptide backbone by up to 8 Å and dramatically closing the cleft between the lobes. The net effect is the appearance that the enzyme molecule has virtually swallowed the glucose molecule, enclosing the substrate glucose in an environment of relative inaccessibility to solvent water and, presumably, pressing it close to the ATP, in the optimum orientation for direct in-line phosphoryl transfer.

REFERENCES

1. J. L. Dahl and L. E. Hokin, *Annu. Rev. Biochem.* **43,** 327 (1974).
2. L. E. Hokin and A. Yoda, *Proc. Natl. Acad. Sci. (USA)* **52,** 454 (1964).
3. D. W. Towle and C. J. Copenhaver, Jr., *Biochim. Biophys. Acta* **203,** 124 (1970).
4. H. Matsui and A. Schwartz, *Biochim. Biophys. Acta* **128,** 380 (1966).
5. C. Hegyvary and R. L. Post, *J. Biol. Chem.* **246,** 5234 (1971).
6. J. C. Scou, *Physiol. Rev.* **45,** 596 (1965).
7. J. S. Britten and M. Blank, *Biochim. Biophys. Acta* **159,** 160 (1968).
8. I. A. Skulskii, V. Manninen, and J. Jarnefelt, *Biochim. Biophys. Acta* **298,** 702 (1973).
9. A. Atkinson, S. Hunt, and A. G. Lowe, *Biochim. Biophys. Acta* **167,** 469 (1968).
10. A. Atkinson and A. G. Lowe, *Biochim. Biophys. Acta* **266,** 103 (1972).
11. R. Rendi and M. L. Uhr, *Biochim. Biophys. Acta* **89,** 520 (1964).
12. G. J. Siegel and B. Goodwin, *J. Biol. Chem.* **247,** 3630 (1972).
13. J. Jensen and J. G. Nørby, *Biochim. Biophys. Acta* **233,** 395 (1971).
14. J. G. Nørby and J. Jensen, *Biochim. Biophys. Acta* **233,** 104 (1971).

15. J. C. Skou, *Biochim. Biophys. Acta* **42,** 6 (1960).
16. R. L. Post, A. K. Sen, and A. S. Rosenthal, *J. Biol. Chem.* **240,** 1437 (1965).
17. S. P. R. Rose, *Nature* **199,** 375 (1963).
18. J. S. Charnock and R. L. Post, *Nature* **199,** 910 (1963).
19. R. W. Albers, S. Fahn, and G. J. Koval, *Proc. Natl. Acad. Sci. (USA)* **50,** 474 (1963).
20. R. Gibbs, P. M. Roddy, and E. Titus, *J. Biol. Chem.* **240,** 2181 (1965).
21. L. E. Hokin, P. S. Sastry, P. R. Galsworthy, and A. Yoda, *Proc. Natl. Acad. Sci. (USA)* **54,** 177 (1965).
22. R. Blostein, *Biochem. Biophys. Res. Commun.* **24,** 598 (1966).
23. T. Kanazawa, M. Saito, and Y. Tonomura, *J. Biochem.* (Tokyo) **61,** 555 (1967).
24. A. K. Sen, T. Tobin, and R. L. Post, *J. Biol. Chem.* **244,** 6596 (1969).
25. A. Kahlenberg, P. R. Galsworthy, and L. E. Hokin, *Arch. Biochem. Biophys.* **126,** 331 (1968).
26. C. Degani and P. D. Boyer, *J. Biol. Chem.* **248,** 8222 (1973).
27. I. Nishigaki, F. T. Chen, and L. E. Hokin, *J. Biol. Chem.* **249,** 4911 (1974).
28. E. T. Wallick, L. K. Lane, and A. Schwartz, *Annu. Rev. Physiol.* **41,** 397 (1979).
29. A. S. Hobbs and R. W. Albers, *Annu. Rev. Biophys. Bioeng.* **9,** 259 (1980).
30. E. T. Fossel, R. L. Post, D. S. O'Hara, and T. W. Smith, *Biochemistry* **20,** 7215 (1981).
31. R. L. Post, G. Toda, and F. N. Rogers, *J. Biol. Chem.* **250,** 691 (1975).
32. A. K. Sen and R. L. Post, *J. Biol. Chem.* **239,** 345 (1964).
33. R. Whittam and M. E. Ager, *Biochem. J.* **97,** 214 (1965).
34. P. J. Garrahan and I. M. Glynn, *J. Physiol.* **192,** 217 (1967).
35. R. C. Thomas, *J. Physiol.* **201,** 495 (1969).
36. S. M. Goldin, *J. Biol. Chem.* **252,** 5630 (1977).
37. S. Uesugi, N. C. Dulak, J. F. Dixon, T. D. Hexum, J. L. Dahl, J. F. Perdue, and L. E. Hokin, *J. Biol. Chem.* **246,** 531 (1971).
38. L. E. Hokin, J. L. Dahl, J. D. Deupree, J. F. Dixon, J. F. Hackney, and J. F. Perdue, *J. Biol. Chem.* **248,** 2593 (1973).
39. J. Kyte, *J. Biol. Chem.* **246,** 4157 (1971).
40. J. Kyte, *J. Biol. Chem.* **247,** 7642 (1972).
41. W. M. Hart and E. O. Titus, *J. Biol. Chem.* **248,** 4674 (1973).
42. J. Kyte, *J. Biol. Chem.* **249,** 3652 (1974).
43. D. E. Metzler, *Biochemistry—The Chemical Reactions of Living Cells,* Academic, New York (1977), p. 272.
44. R. P. Kernan, *Cell Potassium,* Wiley-Interscience, New York (1980).
45. W. F. Harrington, in *The Proteins,* 3rd ed., vol. IV, H. Neurath and R. L. Hill, Eds., Academic, New York (1979), Chapter 3.

46. R. S. Adelstein and E. Eisenberg, *Ann. Rev. Biochem.* **49,** 921 (1980).
47. Y. Tonomura and K. Kanazawa, *J. Biol. Chem.* **240,** PC4110 (1965).
48. R. W. Lymn and E. W. Taylor, *Biochemistry* **9,** 2975 (1970).
49. M. R. Webb and D. R. Trentham, *J. Biol. Chem.* **255,** 8629 (1980).
50. C. R. Bagshaw and D. R. Trentham, *Biochem. J.* **133,** 323 (1973).
51. M. A. Geeves, M. R. Webb, C. F. Midelfort, and D. R. Trentham, *Biochemistry* **19,** 4748 (1980).
52. J. A. Sleep, D. D. Hackney, and P. D. Boyer, *J. Biol. Chem.* **255,** 4094 (1980).
53. M. N. Malik and A. Martonosi, *Arch. Biochem. Biophys.* **144,** 556 (1971).
54. D. R. Trentham, R. G. Bardsley, J. F. Ecclston, and A. G. Weeds, *Biochem. J.* **126,** 635 (1972).
55. R. W. Lymn and E. W. Taylor, *Biochemistry* **10,** 4617 (1971).
56. G. G. Hammes and D. Kovachi, *J. Am. Chem. Soc.* **84,** 2069 (1962).
57. J. Ricard, G. Noat, C. Got, and M. Borel, *Eur. J. Biochem.* **31,** 14 (1972).
58. D. C. Hohnadel and C. Cooper, *Eur. J. Biochem.* **31,** 180 (1972).
59. H. J. Fromm, E. Silverstein, and P. D. Boyer, *J. Biol. Chem.* **239,** 3646 (1964).
60. H. G. Britton and J. B. Clark, *Biochem. J.* **128,** 104P (1972).
61. F. J. Moffet and W. A. Bridger, *J. Biol. Chem.* **245,** 2758 (1970).
62. W. A. Bridger, W. A. Millen, and P. D. Boyer, *Biochemistry* **7,** 3608 (1968).
63. A. Kaji and S. P. Colowick, *J. Biol. Chem.* **240,** 4454 (1965).
64. L. Cheng, T. Inagami, and S. P. Colowick, *Fed. Proc.* **32,** 667 (1973).
64a. J. R. Knowles, *Ann. Rev. Biochem.* **49,** 877 (1980).
65. C. F. Midelfort and I. A. Rose, *J. Biol. Chem.* **251,** 5881 (1976).
66. G. A. Orr, J. Simon, S. R. Jones, G. J. Chin, and J. R. Knowles, *Proc. Natl. Acad. Sci. (USA)* **75,** 2230 (1978).
67. W. A. Blättler and J. R. Knowles, *J. Am. Chem. Soc.* **101,** 510 (1979).
68. G. Lowe and B. V. L. Potter, *Biochem. J.* **199,** 227 (1981).
69. K. D. Wilkinson and I. A. Rose, *J. Biol. Chem.* **254,** 12567 (1979).
70. S. P. Colowick, in *The Enzymes,* vol. IX, P. D. Boyer, Ed., Academic, New York (1973), Chapter 1.
71. R. J. Fletterick, D. J. Bates, and T. A. Steitz, *Proc. Natl. Acad. Sci. (USA)* **72,** 38 (1975).
72. W. S. Bennett, Jr. and T. A. Steitz, *J. Mol. Biol.* **140,** 183 (1980).
73. W. S. Bennett, Jr. and T. A. Steitz, *J. Mol. Biol.* **140,** 211 (1980).
74. M. Shoham and T. A. Steitz, *J. Mol. Biol.* **140,** 1 (1980).
75. W. S. Bennett, Jr. and T. A. Steitz, *Proc. Natl. Acad. Sci. (USA)* **75,** 4848 (1978).
76. A. R. Sielecki, M. N. G. James, and C. G. Broughton, in *Computational Crystallography,* D. Sayre, Ed., Oxford University Press, Oxford, (1981), p. 409.

5

ATP and Metabolic Regulation

Considering the central role that ATP plays in metabolism, superficial considerations might lead to the prediction that ATP should, of itself, be a potent metabolic regulator. For example, metabolic situtations that might lead to a significant depletion of ATP concentration might, in principle, be overcome by relief from allosteric inhibition by ATP of catabolic pathways, with resulting stimulation metabolic flux to replenish the level of this nucleotide. Similarly, one could imagine that in resting tissues with a "fully charged" ATP storage battery, high concentrations of ATP could allosterically reduce catabolic activity and simultaneously stimulate pathways for biosynthesis and growth.

These ideas have been popular over the past 25 years since our concept of the interrelationships of metabolic pathways and their regulation have come to be better understood. It happens, however, at least in eukaryotic organisms, that allosteric regulation by ATP, of itself, is probably relatively unimportant. The underlying reason for this lies in the importance of the compound itself: mechanisms exist to ensure that ATP concentrations do not vary much during metabolic stress, and the small fluctuations that may occur are not sufficient to provide for a sensitive switch to respond effectively to short- or long-term changes in the energy demands of a cell. As can be seen below, ATP does play a collaborative role in regulation of catabolism and anabolism.

MAINTENANCE OF ATP CONCENTRATION

It is clear that various animal tissues have adopted strategies for ensuring that the concentration of ATP remains relatively constant. One of the best techniques currently available for measuring ATP concentrations *in vivo* is ^{31}P-NMR, primarily because it is noninvasive and determinations can be made without disrupting the living cells or subjecting them to extreme environments.[1,2] By this technique, for example, Burt et al.[3] have measured the levels of MgATP and other phosphate metabolites in skeletal muscles of various amphibians, birds, and mammals. They found that the level of MgATP remains virtually constant for several hours. In the case of the excised leg muscle of the Northern frog, MgATP remains essentially constant for seven hours at a value representing 20 percent of the total muscle phosphorus; during that period the level of phosphocreatine drops from over 60 percent to near zero, with a reciprocal increase in the content of inorganic phosphate. Clearly, in the dissected muscle, the reserve of phosphocreatine plays a large role in maintaining the MgATP concentration at its normal value of 3.0 ± 0.2 m*M*. Even more impressive are the data recorded for intact gastrocnemius muscles from this animal: MgATP remains constant for seven hours, but there is only a very small decrease in the content of phosphocreatine. In this living muscle, the ATP replenishment is achieved by mobilization of sugar phosphates, presumably by glycogenolysis.

Tornheim and Lowenstein[4] have suggested another complex mechanism for prevention of wide excursions in the ATP concentration during muscular contraction. This involves the operation of the purine nucleotide cycle (q.v.), which includes the action of adenylate deaminase. Deamination of adenylate to ammonia and inosinate, known to occur in direct proportion to muscular work, would bring about the net conversion of ADP to ATP through readjustment of the adenylate kinase equilibrium:

$$2\text{ADP} \rightleftharpoons \text{ATP} + \text{Adenylate}$$

It has been suggested[5] that the allosteric properties of adenylate deaminase may be appropriate for such a role for this enzyme. Further to this suggestion, Chung and Bridger[6] found that adenylate deaminase from cardiac muscle is exquisitely sensitive to allosteric activation by ADP, much more so than by the widely recognized effects of ATP. They have estimated that at near physiological concentrations of adenylate (ca. 0.1 m*M*), 0.65 m*M* ADP is sufficient for half-maximal activation of the enzyme, whereas approximately 90 m*M* ATP would be required for the same effect. This specificity for ADP for pronounced allosteric activation of cardiac adenylate deaminase is easily reconciled with the physiological role for this enzyme suggested by Tornheim and Lowenstein.[4] The allosteric properties allow for negligible activity at physiological concentrations of adenylate until ATP breakdown has brought about the accumulation of some ADP. The elevated ADP level resulting from prolonged strong contraction of the muscle will activate adenylate deaminase to remove adenylate, followed by readjustment of the adenylate kinase equilibrium (see above) and net conversion of ADP to ATP at the expense of the overall pool of adenine nucleotides.

Since ATP homeostasis is maintained but the concentrations of other adenine nucleotides are not so strictly controlled, Atkinson[7] has suggested that a more sensitive index of the cellular energy economy may be the "energy charge," that is, the mole fraction of ATP in the adenine nucleotide pool (the actual mole fraction of ATP plus half the mole fraction of ADP). Many purified enzymes respond in the anticipated way to the energy charge (see Atkinson[7] and citations therein), particularly those isolated from bacterial sources. The importance of the energy charge as a control feature *in vivo* remains a very debatable point, but since the charge itself has a narrow range in living cells, it is difficult to imagine that regulation by this parameter could be the dominant factor in the control of metabolic flux. [Interesting and provocative reading is contained in the opposite viewpoints expressed by D. E. Atkinson and H. J. Fromm on the significance of energy charge; see the Discussion Forum section of *Trends in Biochemical Sciences* **2,** N198 (1977).]

An important but more subtle component that contributes heavily to cytosolic ATP homeostasis is the ADP, ATP transport system of the mitochondrion. This nucleotide transporter is the most abundant protein in mitochondria and the most abundant membrane protein in many eukaryotic cells.[8] It is responsible for the uptake of ADP generated by energy usage in the cytosolic compartment and its replacement with freshly synthesized ATP from the oxidative phosphorylation machinery of the mitochondrion. It is extremely specific for these two nucleotides; adenylate is not transported and must be phosphorylated in the cytosol by the consecutive action of adenylate kinase. It is of considerable interest that this important transporter is coupled to the membrane potential of the mitochondrion. Thus, in uncoupled mitochondria, the rates of counter transport of ADP and ATP through the membrane are symmetrical. In normal coupled mitochondria, however, the rate of uptake of ATP is only about one-fifth that of ADP, although the concentration of the latter is much less, whereas the rate of export of ATP is five times that of ADP. The driving force provided by the mitochondrial membrane potential results in the exchange of ADP-in for ATP-out over 95 percent of the time, thus generating a considerably higher [ATP]:[ADP] ratio in the cytosol than in the mitochondrial compartment.[8]

PROTEIN KINASES AND METABOLIC CONTROL

One of the most significant discoveries in the history of biochemistry was that of cyclic AMP (cAMP) by Earl Sutherland at Vanderbilt University. Coupled at that time with the demonstration of a protein kinase cascade system for the stimulation of glyogen breakdown by Fischer, Krebs, and their colleagues at the University of Washington, this discovery provided a rationale not only for a "second messenger" to act as an intracellular signal for the presence of a hormone on its exocellular receptor, but also for the cascade of amplification of the signal so that very few primary effector cAMP molecules are required for massive metabolic effects.[9] This prototype example of ATP-driven protein kinase-catalyzed phosphorylation does not really represent control by ATP, of course, since the level of ATP is not a factor in regulating the rate or extent of protein phosphorylation *in vivo*. Nevertheless, the diverse effects of various protein kinases on metabolism, cell division, differentiation, morphology, and so on are staggering in their scope, and a thorough discussion of these effects would require several volumes of this size. As an introduction and to illustrate the importance of protein kinase-mediated effects, some of these are listed in Table 5.1.

REGULATION OF PHOSPHOFRUCTOKINASE ACTIVITY

One of two physiologically irreversible steps in the glycolytic pathway is phosphofructokinase, and it is known that regulation of the catalytic activity of this enzyme plays a very major role in "turning on" glycolysis. The enzyme that operates in the reverse direction, fructose-1,6-bisphosphatase, contributes significantly to the stimulation of gluconeogenesis, and this two-enzyme "switch"—with reciprocal controls on each—provides a classic example of a regulatable metabolic valve that allows carbon atoms to flow by means of glycolysis or gluconeogenesis, depending on the prevailing metabolic and hormonal situation.

Until very recently indeed, although regulation of phosphofructokinase was thought to be relatively complex, the intricacies of its precise regulation have been underestimated. For example, any of the most recent textbooks of biochemistry, in considering regulation of phosphofructokinase, points to the sen-

Table 5.1. Some Systems Affected by Protein Phosphorylation

System	cAMP Dependence	Effect of Phosphorylation	Reference[a]
Glycogen phosphorylase	+	Stimulates glycogen breakdown	9
Glycogen synthetase	+	Inhibits glycogen synthesis	10
Adipocyte lipase	+	Stimulates lypolysis	11
Phosphofructokinase	+	Inhibits by raising substrate K_m and increasing sensitivity to inhibitors[b]	12
Pyruvate kinase	+	Raises K_m for PEP; increases sensitivity to F-1,6-P_2 activation	13
Acetyl–CoA carboxylase	−	Inactivates	14
β-Hydroxy-β-methylglutaryl–CoA reductase	−	Inactivates	15
ATP–citrate lyase	+/−	No effect on activity has been demonstrated	16
Pyruvate dehydrogenase complex	−	Inactivates	17
Tropomyosin	+	Stabilizes head-to-tail overlap	18
Phenylalanine hydroxylase	+?	Stimulates	19
Epidermal growth factor, receptor	−	Promotes cell growth	20
Ornithine decarboxylase	+	Stimulates polyamine production	21
Myosin	+	Regulates actomyosin ATPase	22
Troponin I	−	Stimulates Ca^{2+} exchange with troponin	23
Histone	+/−	Has possible effects on mitosis	24
Protein regulator of maturation-promoting factor, frog oocytes	+	Inhibits differentiation	25
Microtubules	+	Promotes polymerization	26
Rous sarcoma virus *pp*60 substrate	−	Promotes phosphotyrosine formation: multiple effects associated with cell transformation	27

[a]References are given as entries to literature and are not intended to provide comprehensive list.

[b]See text for further discussion of phosphofructokinase effectors.

sitivity of the enzyme to allosteric inhibition by both ATP and by citrate and to the synergism that is exhibited by these two inhibitors. One of the problems that such a simple model for regulation of phosphofrucktokinase presents is that, given the fairly narrow range for the concentration of ATP, the enzyme should be perpetually strongly inhibited. This was made even more clear by the discovery[28,29] that liver phosphofructokinase becomes phosphorylated in glucagon-treated cells by means of cAMP-dependent protein kinase; the phosphorylated enzyme is even more sensitive to ATP inhibition and would be expected to have essentially no residual activity at physiological concentrations of ATP.

There is another important mediator that affects phosphofructokinase activity, however. In fall 1980, at least three groups headed by Hers in Belgium, Pilkis in Tennessee, and Uyeda in Texas had detected an activating factor of low molecular weight that had the capacity to relieve inhibition by ATP and citrate. This factor was identified as fructose-2,6-bisphosphate.[30–32] The effects of fructose-2,6-bisphosphate on the activity of phosphofructokinase are dramatic, indeed: the presence of micromolar concentrations of this compound gives essentially complete relief from ATP and citrate inhibition, and the activation is synergistic with physiological concentrations of adenylate.[32,33] Since it may be argued that phosphofructokinase is otherwise perpetually inhibited by ATP, fructose-2,6-bisphosphate may be viewed as a potent and important regulator of glycolytic flux.

Using hepatocytes that had been isolated from rats that had been treated with various drugs, or from mutant rats lacking phosphorylase kinase, Hue et al.[34] found that fructose-2,6-bisphosphate accumulation *in vivo* may be correlated with the liberation of hexose phosphate precursors. Moreover, they demonstrated that treatment with glucagon suppresses the production of fructose-2,6-bisphosphate, suggesting that the enzyme responsible for its production might itself be under the control of cyclic AMP-dependent protein kinase. We return to this point shortly.

It is a general rule for switch systems such as the phosphofructokinase–fructose-1,6-bisphosphatase couple that factors that activate the forward reaction will have reciprocal inhibiting effects on the reverse. Such is the case for fructose-2,6-bisphosphate, which was shown to be a potent inhibitor of fructose-1,6-bisphosphatase. It may be a reflection on the intensity of the research in this area at the time that two papers bearing the identical title, "Inhibition of fructose-1,6-bisphosphatase by fructose-2,6-bisphosphate," were submitted and published within weeks of each other in spring 1981.[35,36] Thus the presence of this factor should effectively throw the switch for the simultaneous stimulation of glycolysis and inhibition of gluconeogenesis, thus avoiding wasteful futile cycling.

These discoveries prompted a search for new enzyme activities in liver responsible for both the production and destruction of fructose-2,6-bisphosphate, which led to the identification of 6-phosphofructo-2-kinase and fructose-2,6-bisphosphatase, respectively.[37–41] The former enzyme, responsible for production of fructose-2,6-bisphosphate, was found to be regulated by cAMP-dependent phosphorylation, with a resulting inactivation of the enzyme, whereas the degrading enzyme is activated by such phosphorylation. To round out this rather dramatic and intricate story, it has recently been found that these two enzyme activities reside *in a single bifunctional protein.*[42] Thus, when this protein is phosphorylated under the influence of glucagon by means of cAMP-dependent protein kinase, the degrading activity is turned on whereas the synthetic activity is turned off. The dephosphorylated form of the bifunctional enzyme has the reciprocal activity!

REFERENCES

1. R. G. Shulman, T. R. Brown, K. Ugurbil, S. Ogawa, S. M. Cohen, and J. A. den Hollander, *Science* **205,** 160 (1979).
2. C. T. Burt and A. M. Wyrwicz, *Trends Biochem. Sci.* **4,** 244 (1979).
3. C. T. Burt, T. Glonek, and M. Bárány, *J. Biol. Chem.* **251,** 2584 (1976).
4. K. Tornheim and J. M. Lowenstein, *J. Biol. Chem.* **247,** 162 (1972).
5. A. G. Chapman and D. E. Atkinson, *J. Biol. Chem.* **248,** 248 (1973).
6. L. Chung and W. A. Bridger, *FEBS Lett.* **64,** 338 (1976).
7. D. E. Atkinson, *Cellular Energy Metabolism and Its Regulation,* Academic, New York (1977).
8. M. Klingenberg, *Trends Biochem. Sci.* **4,** 249 (1979).
9. R. J. Fletterick and N. B. Madsen, *Ann. Rev. Biochem.* **49,** 31 (1980).
10. C. H. Smith, N. E. Brown, and J. Larner, *Biochim. Biophys. Acta* **242,** 81 (1971).
11. J. D. Corbin, E. M. Reimann, D. A. Walsh, and E. Krebs, *J. Biol. Chem.* **245,** 4849 (1970).
12. J. P. Riou, T. H. Claus, and S. J. Pilkis, *Biochem. Biophys. Res. Commun.* **73,** 591 (1976).
13. P. Ekman, E. Dahlqvist, E. Humble, and L. Engström, *Biochim. Biophys. Acta* **429,** 374 (1976).
14. C. A. Carlson and K.-H. Kim, *J. Biol. Chem.* **248,** 378 (1973).
15. D. M. Gibson and T. S. Ingebritsen, *Life Sci.* **23,** 2649 (1978).

16. N. S. Ranganathan, T. C. Linn, and P. A. Srere, in *Protein Phosphorylation,* vol. 8, *Cold Spring Harbor Conferences on Cell Proliferation,* O. M. Rosen and E. G. Krebs, Eds. (1981), p. 735.
17. T. C. Linn, F. H. Pettit, and L. J. Reed, *Proc. Natl. Acad. Sci. (USA)* **62,** 234 (1969).
18. A. Mak, L. B. Smillie, and M. Bárány, *Proc. Natl. Acad. Sci. (USA)* **75,** 3588 (1969).
19. J. Donlon and S. Kaufman, *Biochem. Biophys. Res. Commun.* **78,** 1011 (1977).
20. S. Cohen, M. Chinkers, and H. Ushiro, in *Protein Phosphorylation,* vol. 8, *Cold Spring Harbor Conferences on Cell Proliferation,* O. M. Rosen and E. G. Krebs, Eds. (1981), p. 801.
21. D. H. Russell and M. K. Maddox, *Adv. Enz. Regulation* **17,** 61 (1979).
22. R. S. Adelstein, M. D. Pato, J. R. Sellers, and M. A. Conti, in *Protein Phosphorylation,* vol. 8, *Cold Spring Harbor Conferences on Cell Proliferation,* O. M. Rosen and E. G. Krebs, Eds. (1981), p. 811.
23. R. J. Solero, S. P. Robertson, J. D. Johnson, and M. J. Holroyde, in *Protein Phosphorylation,* vol. 8, *Cold Spring Harbor Conferences on Cell Proliferation,* O. M. Rosen and E. G. Krebs, Eds. (1981), p. 901.
24. L. R. Gurley, J. A. D'Anna, M. S. Halleck, S. S. Barham, R. A. Walters, J. J. Jett, and R. A. Tobey, in *Protein Phosphorylation,* vol. 8, *Cold Spring Harbor Conferences on Cell Proliferation,* O. M. Rosen and E. G. Krebs, Eds. (1981), p. 10733.
25. J. L. Maller, M. Wu, and J. Gerhart, *Devel. Biol.* **295** (1977).
26. D. L. Purich, B. J. Terry, H. D. White, B. A. Coughlin, T. L. Carr, and D. Kristofferson, in *Protein Phosphorylation,* vol. 8, *Cold Spring Harbor Conferences on Cell Proliferation,* O. M. Rosen and E. G. Krebs, Eds. (1981), p. 1143.
27. T. Hunter, *Trends Biochem. Sci.* **7,** 246 (1982).
28. J. A. Brand and H. D. Söling, *FEBS Lett.* **57,** 163 (1975).
29. J. G. Castano, A. Nieto, and J. E. Felieu, *J. Biol. Chem.* **254,** 5576 (1979).
30. E. Van Schaftingen and H.-G. Hers, *Biochem. Biophys. Res. Comm.* **96,** 1524 (1980).
31. E. Van Schaftingen, L. Hue, and H.-G. Hers, *Biochem. J.* **192,** 897 (1981).
32. S. J. Pilkis, M. R. El-Maghrabi, J. Pilkis, T. H. Claus, and D. A. Cumming, *J. Biol. Chem.* **256,** 3171 (1981).
33. K. Uyeda, E. Furuya, and L. J. Luby, *J. Biol. Chem.* **256,** 8394 (1981).
34. L. Hue, P. F. Blackmore, and J. H. Exton, *J. Biol. Chem.* **256,** 8900 (1981).
35. E. van Schaftingen and H. G. Hers, *Proc. Natl. Acad. Sci. (USA)* **78,** 2861 (1981).
36. S. J. Pilkis, M. R. El-Maghrabi, J. Pilkis, and T. Claus, *J. Biol. Chem.* **256,** 3619 (1981).
37. E. van Schaftingen, D. R. Davies, and H.-G. Hers, *Biochem. Biophys. Res. Commun.* **103,** 362 (1981).

38. M. R. El-Maghrabi, T. H. Claus, J. Pilkis, and S. J. Pilkis, *Proc. Natl. Acad. Sci. (USA)* **79,** 315 (1982).
39. E. Furuya, M. Yokoyama, and K. Uyeda, *Proc. Natl. Acad. Sci. (USA)* **79,** 325 (1982).
40. C. S. Richards, M. Yokoyama, E. Furuya, and K. Uyeda, *Biochem. Biophys. Res. Commun.* **104,** 1073 (1982).
41. E. Furuya, M. Yokoyama, and K. Uyeda, *Biochem. Biophys. Res. Commun.* **105,** 264 (1982).
42. M. R. El-Maghrabi, E. Fox, J. Pilkis, and S. J. Pilkis, *Biochem. Biophys. Res. Commun.* **106,** 794 (1982).

PART II

6

ATP Synthesis

BASIC PATHWAYS OF ATP SYNTHESIS

Three major types of pathway exist for the net synthesis of ATP: (1) purine biosynthesis *de novo* (i.e., from nonpurine precursors); (2) synthesis of adenine nucleotides directly from adenine or from adenosine; and (3) synthesis of adenine nucleotides from other purine bases, nucleosides, and nucleotides. The details of these pathways are considered here, together with what is known about their regulation, relative importance, and coordination. Because the literature on the subject is vast, only selected references are provided; there are several recent reviews of this subject.[1–8]

ATP Synthesis *De Novo*

The process of net synthesis of ATP from nonpurine precursors may be considered to proceed in three phases: (1) purine biosynthesis *de novo* consists of 10 reactions by which the following precursors are converted to inosinate: PP-ribose-P, CO_2, glutamine, glycine, aspartate, 10-formyl H_4-folate, and ATP; (2) the conversion of inosinate to adenylate, a pathway of two reactions requiring aspartate and GTP; and (3) the phosphorylation of adenylate to ADP and finally ATP.

Inosinate Synthesis De Novo

The pathway of inosinate biosynthesis *de novo* consists of 10 reactions, which may be outlined as follows.

Amidophosphoribosyltransferase (EC 2.4.2.14):

$$\text{PP-Ribose-P} + \text{Glutamine} \xrightarrow{Mg^{2+}} \text{Phosphoribosylamine} + \text{Glutamate} + PP_i \tag{6}$$

Phosphoribosyl glycinamide synthetase (EC 6.3.4.13):

$$\text{Phosphoribosylamine} + \text{Glycine} + \text{ATP} \xrightarrow{Mg^{2+}} \text{Phosphoribosyl glycinamide} + \text{ADP} + P_i \tag{7}$$

Phosphoribosyl glycinamide formyltransferase (EC 2.1.2.3):

$$\text{Phosphoribosyl glycinamide} + \text{10-Formyl } H_4\text{-folate} \rightarrow \text{Phosphoribosyl formylglycinamide} + H_4\text{-folate} \tag{8}$$

Phosphoribosyl formylglycinamide synthetase (EC 6.3.5.3):

$$\text{Phosphoribosyl formylglycinamide} + \text{Glutamine} + \text{ATP} \xrightarrow{Mg^{2+}} \text{Phosphoribosyl formylglycinamidine} + \text{Glutamate} + \text{ADP} + P_i \tag{9}$$

Phosphoribosyl aminoimidazole synthetase (EC 6.3.3.1):

$$\text{Phosphoribosyl formylglycinamidine} + \text{ATP} \xrightarrow[K^+]{Mg^{2+}} \text{Phosphoribosyl aminoimidazole} \tag{10}$$

Phosphoribosyl aminoimidazole carboxylase (EC 4.1.1.21):

$$\text{Phosphoribosyl aminoimidazole} + HCO_3^- \rightarrow \text{phosphoribosyl aminoimidazole carboxylate} \tag{11}$$

Phosphoribosyl aminoimidazole succinocarboxamide synthetase (EC 6.3.2.6):

$$\begin{array}{l} \text{Phosphoribosyl} \\ \text{aminoimidazole} \\ \text{carboxylate} \\ + \text{ Aspartate } + \text{ ATP} \end{array} \xrightarrow{Mg^{2+}} \begin{array}{l} \text{Phosphoribosyl} \\ \text{aminoimidazole} \\ \text{succinocarboxamide} \\ + \text{ ADP } + P_i \end{array} \qquad (12)$$

Adenylosuccinate lyase (EC 4.3.2.2):

$$\begin{array}{l} \text{Phosphoribosyl} \\ \text{aminoimidazole} \\ \text{succinocarboxamide} \end{array} \rightarrow \begin{array}{l} \text{Phosphoribosyl} \\ \text{aminoimidazole} \\ \text{carboxamide} \\ + \text{ Fumarate} \end{array} \qquad (13)$$

Phosphoribosyl aminoimidazolecarboxamide formyltransferase (EC 2.1.2.3):

$$\begin{array}{l} \text{Phosphoribosyl} \\ \text{aminoimidazole} \\ \text{carboxamide} \\ + \text{ 10-Formyl } H_4\text{-folate} \end{array} \rightarrow \begin{array}{l} \text{Phosphoribosyl} \\ \text{formamidoimidazole} \\ \text{carboxamide} \\ + \text{ } H_4\text{-folate} \end{array} \qquad (14)$$

IMP cyclohydrolase (EC 3.5.4.10):

$$\begin{array}{l} \text{Phosphoribosyl} \\ \text{formamidoimidazole} \\ \text{carboxamide} \end{array} \rightarrow \text{Inosinate} \qquad (15)$$

Conversion of Inosinate to Adenylate

Inosinate, the first complete purine ribonucleotide formed in the process of ATP synthesis, is next converted to adenylate in the course of two reactions.

Adenylosuccinate synthetase (EC 6.3.4.4):

$$\text{Inosinate} + \text{Aspartate} + \text{GTP} \xrightarrow{Mg^{2+}} \text{Adenylosuccinate} + \text{GDP} + P_i \qquad (16)$$

Adenylosuccinate lyase (EC 4.3.2.2):

$$\text{Adenylosuccinate} \rightarrow \text{Adenylate} + \text{Fumarate} \qquad (17)$$

Phosphorylation of Adenylate

Finally, adenylate is phosphorylated to ADP and ATP by reactions that are considered in Chapter 3.

Several details regarding this pathway should be noted:

1. Phosphoribosylamine, the product of the first reaction, may also be synthesized by using ammonia plus either ribose-5-P or PP-ribose-P. However, the ammonia concentrations required for these alternative reactions are quite high, and there is no reason to believe that they operate under physiological conditions.[2,5]
2. The one-carbon donor in reaction (8) was formerly thought to be 5,10-methenyl H_4-folate. However, in recent years increasing evidence has accumulated suggesting that both of the folate coenzyme reactions in this pathway involve 10-formyl H_4-folate.[9,10]
3. Although several of the individual reactions of the pathway are reversible, the pathway as a whole is irreversible, and there is no reason to believe that individual reactions operate reversibly under normal conditions.[2]
4. One enzyme, adenylosuccinate lyase, catalyzes two separate reactions [(13) and (17)] in this pathway. Other pairs of similar reactions [glutamine amide transfer, reactions (6) and (9); aspartate amino transfer, reactions (12) and (16); one-carbon transfer, reactions (8) and (14); and aromatic ring closure, reactions (10) and (15)] are catalyzed by distinct enzymes.

Formation of Adenylate From Adenine and Adenosine

Both the purine base adenine and the purine ribonucleoside adenosine can be converted to adenylate by distinct one-step reactions.

Adenine phosphoribosyltransferase (EC 2.4.27):

$$\text{Adenine} + \text{PP-ribose-P} \xrightarrow{Mg^{2+}} \text{Adenylate} + PP_i \tag{18}$$

Adenosine kinase (EC 2.7.1.20):

$$\text{Adenosine} + \text{ATP} \xrightarrow{Mg^{2+}} \text{Adenylate} + \text{ADP} \tag{19}$$

Thereafter, the product adenylate is readily phosphorylated to ADP and then to ATP. The purine substrates of these reactions may be provided exogenously and synthesized endogenously.

Alternative Pathways of Inosinate Synthesis

Purine bases, ribonucleosides, and ribonucleotides other than adenosine and adenine may at least potentially be precursors of ATP synthesis through conversion first to inosinate. Two reactions are of primary importance.

Hypoxanthine phosphoribosyltransferase (EC 2.4.2.8):

$$\text{Hypoxanthine} + \text{PP-ribose-P} \xrightarrow{Mg^{2+}} \text{Inosinate} + PP_i \tag{20}$$

Guanylate reductase (EC 1.6.6.8):

$$\text{Guanylate} + \text{NADPH} \rightarrow \text{Inosinate} + NH_3 + \text{NADP} \tag{21}$$

The variety of potential sources of hypoxanthine and guanylate should be mentioned briefly. Thus inosine is a precursor of hypoxanthine, and it, in turn, can be formed both from inosinate and from adenosine. Guanylate can be formed by the dephosphorylation of GDP and GTP, from inosinate by way of xanthylate, and from guanine by the guanine phosphoribosyltransferase (EC 2.4.2.8) reaction. In some cells xanthine is converted to xanthylate (usually at low rates), but in many systems it cannot be utilized.

The overall scheme of ATP synthesis may be pictured as follows:

$$\begin{array}{ccccccccc} \text{ATP} & \leftarrow & \text{ADP} & \leftarrow & \text{Adenylate} & \leftarrow\leftarrow & \text{Inosinate} & \leftarrow & \textit{De novo} \\ & & & & \uparrow \quad \uparrow & & \uparrow \quad \uparrow & & \\ & & & & \text{Adenosine} \quad \text{Adenine} & & \text{Hypoxanthine} \quad \text{Guanylate} & & \end{array} \tag{22}$$

REGULATION OF ATP SYNTHESIS AS A WHOLE

It does not suffice to simply examine the individual pathways of ATP synthesis one by one; to provide a complete picture of this process it is also necessary to consider these pathways together. Clearly, in intact cells and tissues they operate together in a coordinated and integrated manner to meet the needs of cells for ATP (and nucleic acids and other nucleotides). Three issues are of special importance: (1) what the various functions of the biosynthetic pathways under consideration here are; (2) how the rates of these alternative pathways might be coordinated; and (3) what the relative rates of the several alternative pathways of ATP synthesis are in different cells and tissues and under different conditions.

Alternative Functions of the Biosynthetic Pathways

The pathways of ATP synthesis have several functions in addition to the production simply of ATP; these should be mentioned here at least briefly.

The De Novo *Pathway*

The first part of this pathway, from phosphoribosylamine to inosinate, is also part of the process by which GTP is produced *de novo;* in addition, GTP and ATP both are precursors of several important coenzymes and of RNA, and ADP and GDP are precursors of deoxyribonucleotides and hence of DNA. Thus the total amount of purine nucleotides synthesized *de novo* is distributed to a number of products in addition to ATP.

In birds and some insects and reptiles, the *de novo* pathway of inosinate synthesis plays an essential role in nitrogen excretion. In these organisms the end product of protein nitrogen metabolism is uric acid, produced by use of the nitrogens contributed through glutamine, glycine, and aspartate; the inosinate that is the immediate end product of the pathway is then catabolized to produce uric acid. In such uricotelic organisms, most of the inosinate formed *de novo* is thus converted to uric acid, whereas only what is needed is metabolized further to ATP; clearly, there is tight control over the alternative pathways of inosinate metabolism. There is no evidence that ATP concentrations in tissues (especially the livers) of uricotelic animals are higher than those of ureotelic organisms; however, this question has not been thoroughly investigated.

Utilization of Adenine and Adenosine

The synthesis of adenylate, ADP, and ATP from adenine and adenosine may not only contribute to the net synthesis of ATP, but may also participate in metabolic cycles in which ATP is both utilized and resynthesized. This topic is considered in Chapter 9.

Inosinate Synthesis From Hypoxanthine

As discussed above, the synthesis of inosinate from hypoxanthine may not only lead to the net synthesis of ATP, but may also be part of cycles of turnover of nucleotides and reutilization of the catabolites thus produced. In this case, inosinate and other nucleotides formed from RNA and DNA may be sources of hypoxanthine, as well as of ATP itself. Again, this point is considered further in Chapter 9.

Inosinate Synthesis From Guanylate

This process not only has a synthetic function with respect to ATP (which is minor), but also may participate in the catabolism of GTP and other guanine nucleotides. Even in this case, however, there are important variations among species.[11]

Coordination of Alternative Pathways

The several pathways of ATP synthesis are not only linked by having the same product, but also, in some cases, by sharing common enzymes, substrates, or regulatory metabolites.

One enzyme, adenylosuccinate lyase, participates both in the *de novo* synthesis of inosinate and in the conversion of inosinate to adenylate. It thus plays a role in adenylate synthesis from hypoxanthine and guanylate as well as in the *de novo* pathway. However, in animal cells, at least, adenylosuccinate synthetase is usually thought to be the rate-limiting step in the conversion of inosinate to adenylate, and the multiple role of adenylosuccinate lyase is not known to be of regulatory importance.

The substrate aspartate participates in both the *de novo* synthesis of inosinate and the conversion of inosinate to adenylate.

A very important coordinating factor is the substrate PP-ribose-P. It is required for three of the pathways considered here: inosinate synthesis *de novo;* adenine phosphoribosyltransferase; and hypoxanthine phosphoribosyltransferase.

Finally, adenylate, ADP, and ATP act as regulatory metabolites with respect to several of the pathways of ATP synthesis.

All these potential regulatory factors are considered in greater detail in the following pages.

Relative Rates and Rate-Limiting Steps

What is the relative rate and significance for cells of each of the five routes of ATP biosynthesis listed above? How do their relative rates differ among cell types, and with experimental conditions?

Methodological Considerations

Although it would be highly desirable to assess the relative importance of the several alternative pathways of ATP synthesis, this in fact is extremely difficult to do; published estimates are to be viewed with great caution.[12] Among the methodological obstacles are the following.

First, methods are only beginning to be developed for the measurement of the rates of the several cyclic pathways mentioned above (e.g., those involving adenine phosphoribosyltransferase, adenosine kinase, and hypoxanthine phosphoribosyltransferase), and at present it is not possible to distinguish between their contribution to the net synthesis of ATP and their cyclic function.

Second, inasmuch as the normal concentrations and rates of synthesis of the purine base and nucleoside substrates for ATP synthesis (principally adenine, adenosine, and hypoxanthine) are not accurately known in most systems, it is not possible to provide radioactive precursors in ways that will duplicate normal conditions in cells or tissues. Thus in many studies, relatively high concentrations of radioactive precursors have been used, and some estimate of the maximal rates of the pathways is obtained; what relationship these values have to the actual physiological rates is not clear. Even the real significance of rates measured at tracer concentrations of precursors is not certain, and in these studies, corrections for changes in specific activity due to dilution by endogenous substrate probably should be made.

Third, measurement of the rate of *de novo* synthesis is commonly carried out by using radioactive glycine or formate. However, only a few investigators determine the specific activity of these precursors in the cells and tissues studied during the course of the experiments,[13] and without this information, absolute rates of synthesis cannot be determined.

Fourth, in the use of genetic or pharmacological lesions to block one or another of the pathways of ATP synthesis, the possibility must always be considered that the rates of other pathways may change in a compensatory manner.

In no case have all these difficulties been overcome.

Rates of Alternative Pathways

Although the methodological problems outlined above make it difficult to assess precisely the actual rates of each of the major pathways of ATP synthesis (*de novo,* from adenine and from adenosine) either in "normal" cells or when these rates are perturbed, certain extreme situations can be defined.

Whole animals, for example, do not require purines in their diet and hence are able to depend entirely on *de novo* synthesis to furnish ATP and other purines. Within the body, however, not all cells need be able to synthesize ATP *de novo* because purines can also be acquired via the circulation (see Chapter 8). The mature erythrocyte is one type of cell that totally lacks the *de novo* pathway.[2,14]

Most cultured mammalian cells also are able to grow in the absence of exogenous purines, although growth sometimes is stimulated by their addition. Cultured cells lacking adenine phosphoribosyltransferase, adenosine kinase, or hypoxanthine phosphoribosyltransferase grow perfectly normally on the basis of purine biosynthesis *de novo*. However, the *de novo* pathway can also be blocked (genetically or with drugs), and in such cases cell growth can quite adequately be supported by adenine, adenosine, or hypoxanthine.

Most mammalian cells, however, possess all three major routes of ATP synthesis, although their apparent rates vary widely among cell types. Judgments sometimes are made that one cell type has a slow rate of one or another pathway and another cell type has a fast rate, but such conclusions need to be severely qualified not only on methodological grounds, but also because the *needs* of cells for ATP synthesis quite likely vary widely as well. At the present time it is necessary to admit that there probably are few reliable data as to the relative rates of these processes.

However, a few examples may be given of the relative rates of adenine nucleotide synthesis from various radioactive precursors, when these have been compared in a single study. Thus in chopped mouse brain *in vitro,* relative rates were 3, 20, 13, and 45 when glycine, adenine, hypoxanthine, and adenosine, respectively, were precursors. When guanosine was used in place of glucose as a source of ribose for PP-ribose-P, nucleotide synthesis from adenine and hypoxanthine both increased by approximately 2.3-fold.[15] In another study, radioactive precursors were perfused into isolated rat hearts; relative rates of nucleotide formation from adenosine, adenine, inosine, and hypoxanthine were 177, 23, 30, and 27, respectively.[16]

An entirely different approach was taken in a study of alternative pathways of ATP synthesis in Ehrlich ascites tumor cells *in vivo*. Here specific or relatively specific inhibitors of individual pathways were administered, and the relative importance of individual pathways was calculated from the extent to which ATP concentrations decreased as a result of drug treatment. These results suggested that *de novo* synthesis is responsible for 85 percent of net ATP synthesis, hypoxanthine phosphoribosyltransferase and adenine phosphoribosyltransferase for about 7 percent each, and guanylate reductase for less than 1 percent.[17] The assumptions involved in this approach are complex and are discussed in full in the original paper.

Even considering the limitations outlined above, however, it is possible to conclude that one or another of the pathways of ATP synthesis changes in response to various stimuli. Phytohemagglutinin-stimulated blastogenesis of human peripheral lymphocytes, for example, is accompanied by increased rates of PP-ribose-P synthesis and of nucleotide synthesis *de novo* and from adenine, hypoxanthine, and adenosine.[18–20] To give another example, purine biosynthesis *de novo* is greatly accelerated during the process of cardiac hypertrophy in the rat, and the determining factor is believed to be the provision of PP-ribose-P for this process.[21,22] Finally, the incorporation of adenosine into adenine nucleotides is increased when cultured mammalian cells are deprived of serum, whereas the rates of other pathways decrease.[23]

Individual Enzymes of ATP Synthesis

Among the factors that may regulate the reactions of ATP biosynthesis are the amount, localization, and properties of the individual enzymes and their regulation by products and other metabolites. Therefore, selected data for the key enzymes of the alternative pathways of ATP synthesis are considered next.

PP-Ribose-P Synthetase

Although this enzyme is not, strictly speaking, part of the individual pathways of ATP synthesis described above, it should be considered because it provides the substrate for three of the biosynthetic pathways (inosinate synthesis *de novo*, adenine phosphoribosyltransferase, hypoxanthine phosphoribosyltransferase) and because it is one site of end-product regulation of ATP synthesis.

The apparent Michaelis constants for ribose–5-P are in the range 33–290 μM[24] and for ATP, 14–220 μM. Inhibitors of mammalian PP-ribose-P include a variety of purine ribonucleotides, 2,3-diphosphoglycerate, and the reaction products PP-ribose-P and adenylate; ADP is the most potent inhibitor (K_i, 10 μM). PP-Ribose-P synthetase activity has an absolute requirement for P_i; maximal concentrations are in the range 10–100 m*M*, well above the physiological range 0.5–1.5 m*M*.

PP-Ribose-P synthetase is composed of a basic subunit of molecular weight 33,200 that undergoes reversible aggregation to forms containing 2, 4, 8, 16, and 32 such subunits.[25–27] The extent of aggregation is dependent on enzyme concentration and the concentration of any of a number of ligands, including substrates, products, and inhibitors. It is the largest aggregates (16 and 32 subunits) that are catalytically active, whereas the smaller units have little activity.

Amidophosphoribosyltransferase

This, the first enzyme of the pathway of ATP synthesis *de novo,* catalyzes what appears to be the rate-limiting step in the pathway as far as inosinate. It has been characterized in a number of microbial, avian, and mammalian cells or tissues and has been highly purified from several sources.[2,5] Apparent Michaelis constants are in the range 60–900 μM for PP-ribose-P and 1.0–1.7 m*M* for glutamine; the relationship between PP-ribose-P concentrations and velocity is sigmoidal rather than hyperbolic.

Most isolated amidophosphoribosyltransferases are inhibited by adenine and guanine nucleotides, but the kinetics of inhibition vary from simple competitive inhibition with respect to PP-ribose-P, to more complex allosteric interactions, depending on the source of enzyme. One or more types of regulatory nucleotide binding site also appear to be present, again depending on the preparation.

Physically, two forms of the enzyme have been identified, with molecular weights of 130,000 and 270,000. The small form is converted to the large form by incubation with purine ribonucleotides, and the large form is converted to the small form by incubation with PP-ribose-P. Overall catalytic activity is correlated with the amount of enzyme in the small form.[28]

Other Enzymes of De Novo *Synthesis*

The nine enzymes of inosinate synthesis *de novo* following amidophosphoribosyltransferase seldom play important regulatory roles and thus are not considered further; for basic details, see Henderson[2]).

Adenylosuccinate Synthetase and Adenylosuccinate Lyase

The metabolism of inosinate by its three alternative pathways (to adenylate, to xanthylate and guanine nucleotides, and to inosine) is an important point at which ATP biosynthesis *de novo* is regulated; this is discussed further below. The formation and cleavage of adenylosuccinate from inosinate is not only an essential part of the overall pathway, therefore, but also a site of regulation.[7]

There is only limited information regarding the activities of adenylosuccinate synthetase and adenylosuccinate lyase in different cell types and among species. In the rat, the relative activities of the synthetase in soleus muscle, "skeletal" muscle, and brain and kidney cortex are 0.72, 0.46, 0.036, and 1.27 units, respectively. In the latter three tissues, the activities of the lyase are 0.38, 0.18, and 1.38 units.[29–31]

The Michaelis constant of the synthetase for inosinate is about 30 μM, and that of the lyase for adenylosuccinate is approximately 10 μM. As the concentration of inosinate in Ehrlich ascites tumor cells and cultured cell lines is about 30 μM and that of adenylosuccinate approximately 4 μM, it can be concluded that the rate of both reactions depends greatly on the concentrations of these substrates.[7]

The enzyme is inhibited by adenylosuccinate and adenylate and by guanylate and GDP; K_i values are in the range 10–170 μM.[32] Neither elevation nor lowering of ATP or GTP concentrations had any demonstrable effect on the apparent activity of adenylosuccinate synthetase in intact Ehrlich ascites tumor cells.[33,34]

Adenine Phosphoribosyltransferase

Studies of this enzyme have been well reviewed.[1,6,35] It has been purified from a number of sources, and its tissue distribution has been determined. Michaelis constants for adenine are in the range 1–125 μM; those for PP-ribose-P are in the range 5–37 μM. Adenine phosphoribosyltransferase is inhibited by the immediate product, adenylate, and also by ADP, ATP, and guanine nucleotides.

Adenosine Kinase

This enzyme has been partially purified from yeast, rabbit liver, Ehrlich ascites tumor cells, and rat heart[8] and, more recently, highly purified from rabbit liver[36] and human placenta.[37] The Michaelis constants for adenosine are in the range 0.4–1.6 μM; those for ATP are in the range 10–200 μM. Adenosine kinase has been found to be inhibited at high concentrations of adenosine, with 50 percent inhibition obtained at approximately 100 μM.[38–40] Adenosine kinase is also subject to product inhibition, with a K_i for ADP of 60 μM.[41]

Role of Substrate Concentration

The pathways of ATP synthesis *de novo* and from adenine and adenosine are significantly regulated by the concentrations of the required substrates of these pathways. Three types of factor are important in determining the availability of one or another of these substrates: rate of intracellular synthesis; alternative routes of metabolism; and exogenous supply.

PP-Ribose-P

The synthesis of PP-ribose-P requires first the formation of ribose-5-P; this, in turn, can be made from hexoses by means of the oxidative and nonoxidative pentose phosphate pathways, from pentoses, and at least potentially by gluconeogenesis leading to glycolytic intermediates that can be converted to ribose-5-P. In the second step, ribose-5-P reacts with ATP to form PP-ribose-P + adenylate; this is the PP-ribose-P synthetase reaction and has an absolute requirement for P_i as an allosteric activator.[24,42]

The dependence of PP-ribose-P synthesis and of purine biosynthesis *de novo* on glucose and on factors that influence the rates of the oxidative and nonoxidative pentose phosphate pathways has been demonstrated in many biological systems; this has been reviewed in detail by Henderson.[2]

As most mammalian tissues do not phosphorylate and metabolize pentoses well, it formerly was thought that these sugars could not be used for PP-ribose-P synthesis. More recently, however, the administration of both ribose and xylitol have been shown to very markedly stimulate purine biosynthesis *de novo* in rat heart and skeletal muscle *in vivo*, whereas these sugars were used for nucleotide synthesis in liver and kidney only to a small degree; ribose was a better precursor of PP-ribose-P than glucose in heart and skeletal muscle.[43] It was also found that ribose, ribitol, xylose, and xylitol were all equally effective.

It has not been demonstrated that gluconeogenesis can lead to PP-ribose-P synthesis.

Once formed, ribose-5-P may be converted to PP-ribose-P, or alternatively, be metabolized to ribose-1-P or to glycolytic intermediates by means of the nonoxidative pentose phosphate pathway. In Ehrlich ascites tumor cells incubated with glucose, the ratio of PP-ribose-P to ribose-1-P was 3.5 when cells were incubated in a P_i-free medium or with 1 mM P_i but was 18 with 25 μM P_i.[44] When ribose-5-P is synthesized by phosphorolysis of a ribonucleoside such as inosine (inosine + P_i → ribose-1-P → ribose-5-P), the amount of ribose-5-P that is converted to glycolytic intermediates can be estimated from the amount of lactate that accumulates (when tumor cells are used). In a typical experiment, the concentrations of ribose-1-P, PP-ribose-P, and lactate following incubation with inosine were 0.16, 0.3, and 20 μmol per milliliter of cells, respectively;[45] thus only a fraction of the ribose-5-P formed is converted to PP-ribose-P. In other experiments, the concentrations of ribose-5-P were about twice those of ribose-1-P when cells were incubated with glucose.[44]

As already indicated briefly, PP-ribose-P synthesis in cells is highly dependent on the concentration of P_i; this has been demonstrated in a large number of experimental systems.[2,24,42] Although this stimulatory effect is believed to be due primarily to the effect of P_i on PP-ribose-P synthetase, P_i also stimulates glucose uptake, glycolysis, and the accumulation of ribose-5-P.[44,45]

One factor that affects the availability of PP-ribose-P for ATP synthesis is its utilization by other reactions, including pyrimidine biosynthesis. For example, this process can be considerably accelerated in liver *in vivo* and in some cultured cells by administration of orotate, and under these conditions purine biosynthesis *de novo* is partially inhibited, apparently as a result of depletion of PP-ribose-P.[2] In rat liver preparations *in vitro,* high concentrations of ammonia can interfere with purine nucleotide biosynthesis by stimulating carbamyl-P synthesis and hence pyrimidine biosynthesis *de novo;* again, the availability of PP-ribose-P would be diminished.[46] The relative use of PP-ribose-P by the individual pathways of ATP synthesis is discussed below.

Pathological influences and effects of drugs on PP-ribose-P metabolism are considered in Chapters 10 and 11.

Glutamine

This amino acid is limiting for purine biosynthesis *de novo* under certain conditions, although it does not seem to be as important a regulating factor as PP-ribose-P. However, it, too, is substrate for the first enzyme of the pathway, and hence both substrates can be limiting for the amidophosphoribosyltransferase reaction and for the pathway as a whole.

Purine biosynthesis has been stimulated in many studies by the addition of glutamine to cells incubated *in vitro*.[2] As media used to support the growth of cultured mammalian cells contain high concentrations of this amino acid, however, it is almost never limiting in these systems. In intact animals, it was shown some years ago that cortisone treatment of adrenalectomized rats leads to increased purine biosynthesis and that this probably is due to increased glutamine synthesis secondary to the primary rise in the concentration of glutamate and probably also of ammonia. One result of increased purine biosynthesis was a 60 percent increase in liver ATP concentrations compared with adrenalectomized controls.[47,48] It has also been shown that the intraperitoneal injection of glutamine into mice stimulates purine biosynthesis *de novo* in liver and kidney and, in separate experiments, in Ehrlich ascites tumor cells growing in the peritoneal cavity.[49]

Aspartate

Although aspartate concentrations are rarely considered to be rate limiting for inosinate synthesis *de novo,* this has been demonstrated by using cells incubated *in vitro*.[49] In addition, there is clear evidence that this amino acid can influence the rate of the adenylosuccinate synthetase reaction. Thus the addition of aspartate to Ehrlich ascites tumor cells incubated in amino acid-free media considerably increased the conversion of inosinate to adenine nucleotides.[50] Asparagine can replace aspartate, apparently because of intracellular deamidation; even the addition of glutamine alone stimulates the adenylosuccinate synthetase reaction as a result of its conversion to aspartate in cells. Addition of aspartate, through its stimulation of the adenylosuccinate synthetase reaction, also decreases the extent of inosinate breakdown to inosine.

Guanosine Triphosphate

This nucleotide is also a substrate of adenylosuccinate synthetase. Lowering of GTP concentrations (to ca. 10 percent of normal) in Ehrlich ascites tumor cells by treatment with mycophenolic acid did cause a small (11 percent) apparent inhibition of adenylosuccinate synthetase activity and a 17 percent decrease in adenine nucleotide concentrations.[17] However, lowering of GTP concentrations to less than 10 percent of control in cultured lymphoma L5178Y cells[51] and to approximately 30 percent of normal in cultured neuroblastoma cells[52] did not produce any appreciable change in ATP concentration. Thus this nucleotide did not appear to be limiting for the conversion of inosinate to adenylate in intact cells.

Other Substrates of De Novo *Synthesis*

Although glycine, aspartate, bicarbonate, and 10-formyl tetrahydrofolate have been shown to become limiting for ATP synthesis *de novo* in certain biological systems and under certain conditions,[2] these situations either are rare or are exhibited under pathological conditions or under the influence of drugs; the latter cases are considered in Chapters 9 and 10.

Adenine

Adenine is metabolized only by adenine phosphoribosyltransferase in most systems; hence alternative pathways need not be considered. The rate of synthesis of adenine appears to be relatively low and occurs as part of the process of S-adenosylmethionine utilization for polyamine synthesis (see Chapter 8). Thus the normal rate of adenylate formation from endogenous adenine may be low and probably does not contribute to any net synthesis of ATP.

In contrast, the effective activity of adenine phosphoribosyltransferase can be very considerably accelerated by supplying adenine exogenously, and under these conditions ATP concentrations can be elevated. Thus in studies with Ehrlich ascites tumor cells, ATP concentrations have been increased by 575 nmol per 10^6 cells,[33] 825 nmol,[53] 2100 nmol,[54] and about 2600 nmol;[55] control ATP concentrations were 2500–3000 nmol in these experiments. When human erythrocytes were incubated with adenine under optimal conditions, ATP concentrations rose from 2.5 to 7.0[56] and from 23 to 47[57] units.

Inhibition by ATP of nucleotide formation from adenine in intact cells can be demonstrated, however. Thus when Ehrlich ascites tumor cells were incubated with adenine to elevate ATP concentration 1.2-fold and the unused adenine then removed, subsequent conversion of radioactive adenine to nucleotides was inhibited 45 percent.[33] Whether this inhibition was exerted at the level of PP-ribose-P synthetase, on adenine phosphoribosyltransferase, or both could not be distinguished. Conversely, nucleotide synthesis from adenine was increased in cells that contained lowered concentrations of ATP.[34]

Adenosine

This substrate is metabolized by both phosphorylation and deamination, and the relative rates of these two processes vary with cell type and species and also with adenosine concentration.[8] As is discussed in more detail in Chapter 8, in some cells deamination predominates at all adenosine concentrations, whereas in others phosphorylation is the major pathway at low adenosine concentrations and deamination at high concentrations; in the cases thus far investigated, the latter pattern appears to be the more common. Furthermore, at concentrations of adenosine that are saturating for both processes, this ratio also varies with species. For example, the ratio of phosphorylation to deamination in mouse erythrocytes is 0.024, whereas in mule deer erythrocytes it is 1.22.[8]

Adenosine can potentially be synthesized by means of a number of different pathways, including dephosphorylation of 5′-adenylate, from S-adenosylhomocysteine formed in the process of methylation, and from RNA. The RNA could be derived from the diet, produced by cell death and turnover, or intracellular RNA could be cleaved in the course of normal function and metabolism. At the present time, neither the total combined rate nor the relative importance of these individual sources and reactions is known; the cycle of formation and reutilization of adenosine in relation to the process of methylation is given further attention in Chapter 9. In any case, it would appear that much of the adenosine generated intracellularly is involved in cyclic processes of one kind or another, and its reutilization probably does not lead to a net increase in ATP production.

As in the case of adenine, however, the provision of adenosine exogenously can lead to a marked elevation of ATP concentrations; in many recent studies, adenosine deaminase activity is first blocked with an appropriate inhibitor (see Chapters 8 and 11). Thus in studies using rat hepatocytes, ATP concentrations increased 416 percent following incubation with adenosine;[58] in human erythrocytes, it increased 25 percent;[59] and in rat, dog, and rabbit hearts *in vivo,* ATP concentrations rose between 21 and 48 percent following infusion with adenosine.[60] Incubation of cultured rat heart cells and cultured skeletal muscle cells with adenosine increased ATP concentrations approximately 30 percent.[61]

Alternative Pathways

The metabolism by alternative pathways of any intermediate of the various synthetic routes, could diminish the net amount of ATP produced.

De Novo Intermediates

So far as is known, none of the nonpurine intermediates of the pathway of purine biosynthetic pathway [reactions (6)–(14)] are metabolized by alternative reactions in mammalian cells. The only branch that is known along this pathway is the diversion of phosphoribosyl aminoimidazole for the synthesis of the pyrimidine moiety of thiamine in plant and microbial cells that make this vitamin.[2] Concentrations of thiamine are very low, however, and this should not be an appreciable drain on this pathway even in these systems.

Adenylate

Adenylate, whether produced *de novo,* from adenine or from adenosine, at least potentially may be dephosphorylated and deaminated as well as phosphorylated to ADP and ATP. Both alternative pathways are considered in more detail when they function in catabolic routes (Chapter 7) and deamination, again, in the context of cyclic pathways (Chapter 8). Here, however, the question is the net loss of adenylate by alternative pathways under conditions of ATP synthesis.

Certainly, phosphorylation is the predominant pathway when Ehrlich ascites tumor cells are incubated with radioactive adenine; for example, the relative net rates of phosphorylation, deamination, and dephosphorylation are 2800:40:2, respectively.[8] In tumor cells incubated with radioactive adenosine, the relative net rates of phosphorylation and deamination are 2800:60.[8] The relative rates of these pathways remain to be determined using other biological systems.

Inosinate

The metabolism of inosinate by alternative pathways is of major importance in the regulation of ATP synthesis *de novo*. Two such pathways exist: dephosphorylation to inosine and oxidation to xanthylate with its subsequent conversion to guanine nucleotides. In some cells, inosinate may also be phosphorylated to inosine triphosphate; however, this compound is then readily converted back to inosinate.[62]

There is ample evidence that when the rate of inosinate synthesis increases, much of the "excess" inosinate is dephosphorylated to inosine rather than converted to adenine nucleotides. This was first discovered in gouty human patients who produced purines *de novo* at accelerated rates.[4,63] In normal individuals, label introduced as glycine, for example, is excreted as urinary uric acid relatively slowly because it mixes with ATP and RNA adenine pools and only then is excreted as a result of the turnover of these compounds. In patients who overproduce purines, however, a substantial amount of label is excreted quite rapidly, indicating that some inosinate that is formed is immediately dephosphorylated and converted to uric acid.

The same phenomenon has also been shown in experimental systems. For example, in cultured human skin fibroblasts incubated with radioactive formate, the rate of inosinate synthesis *de novo* can be varied over a 22-fold range by adjusting the concentration of P_i in the medium. Even at the lowest rate of this process, 22 percent of the total inosinate synthesized was excreted into the medium as hypoxanthine (formed from inosine); when the rate of inosinate synthesis doubled, this figure increased to 29 percent. At the highest rate of *de novo* synthesis, approximately 60 percent of the total inosinate synthesized was degraded and excreted.[64]

The other alternative pathway of inosinate metabolism is conversion to xanthylate and guanine nucleotides. This, too, increases with the rate of inosinate synthesis *de novo,* but to a smaller extent, and it soon becomes saturated. Thus in the same study using cultured skin fibroblasts,[64] guanine nucleotide synthesis increased only 2.6-fold as total inosinate synthesis increased 22-fold. Furthermore, guanine nucleotide synthesis appeared to become maximal after only a four-fold increase in rate of inosinate synthesis. In contrast, adenine nucleotide synthesis could be stimulated 15-fold as total nucleotide production increased 22-fold.

The diversion of inosinate to xanthylate and guanine nucleotides would be expected to be affected by end-product regulation of inosinate dehydrogenase. Thus elevated GTP concentrations in Ehrlich ascites tumor cells very markedly inhibited the effective activity of this enzyme;[33] conversely, this enzyme was stimulated when GTP concentrations were reduced.[34] Inosinate dehydrogenase activity was not affected by changes in ATP concentrations.

The relative rate of conversion of inosinate to adenylate and to guanine nucleotides varies among cell types. For example, this ratio was 3:2 in cultured human fibroblasts,[65] 9:1 in chopped preparations of mouse brain,[15] 1:1 in log-phase cultured human lymphoblasts,[66] and 3:1 in Ehrlich ascites tumor cells;[50] for each system, this ratio also varied with the rate of inosinate synthesis.

Regulatory Mechanisms in Intact Cells

The regulation in intact cells of ATP synthesis *de novo,* from adenine and from adenosine, may be considered in two ways: the regulation of each pathway separately and the effect of changes in the rate of one pathway on the rates of the other two.

Response to Substrate Concentrations

All three pathways of ATP synthesis are dependent on the concentration of key substrates, and their rates can be accelerated very greatly over normal if sufficient substrate is provided. This is clearly illustrated by studies of the kinetics of amidophosphoribosyltransferase in intact tumor cells in vitro.[67] Although the apparent Michaelis constant for PP-ribose-P of 1 mM was similar to that obtained from studies of the partially purified enzyme, this is 100–500 times the PP-ribose-P concentrations that are found in cells.[68]

To give another example, purine biosynthesis *de novo* in human patients can be increased as much as thirtyfold above normal;[69] as reported above, incubation with adenine can lead to 100–280 percent increases in ATP concentration, and as much as 400 percent increases were observed in experiments by use of adenosine. These results are most simply interpreted as indicating that under conditions in which substrate concentrations "push" the reactions in the direction of ATP synthesis, regulation by end products is minimal or absent all together.

This conclusion, however, does not appear to be compatible with observations that in cells with elevated concentrations of ATP, subsequent incubation in the absence of adenine shows a marked inhibition of PP-ribose-P synthesis by nucleotide endproducts.[55] One possible explanation is based on observations that PP-ribose-P synthesis from glucose through the oxidative pentose phosphate pathway actually is stimulated in the presence of adenine.[55,70] Thus it would appear that this acceleration of PP-ribose-P synthesis when it is being used for rapid nucleotide formation from adenine more than overcomes any inhibition of PP-ribose-P by the increasing elevation of ATP concentrations.

Nucleotide synthesis from adenosine has fewer potential controls and appears to depend greatly on the availability of adenosine; in addition, however, this process may be limited by adenosine kinase capacity in some cells and may be inhibited at high adenosine concentrations.[40] Phosphorylation of adenosine is also promoted by P_i.[40]

In conclusion, it would appear that high substrate concentrations can overcome any regulation of these biosynthetic pathways by end products.

Interrelationships Among Pathways

Such relationships may be considered through the question: Does a change in the rate of one pathway of ATP synthesis affect the rates of the others? To take these situations one at a time, it must be admitted that there simply is no information regarding the effect of accelerated purine biosynthesis on the rates of nucleotide synthesis from adenine and adenosine. One might conjecture that since the endogenous rates of these routes probably are quite low, they may not be affected; furthermore, the affinity of adenine phosphoribosyltransferase for PP-ribose-P is greater than that of amidophosphoribosyltransferase. Conversely, because these rates are low, they might be more sensitive to perturbation by accelerated *de novo* synthesis and elevated ATP concentrations.

The effect of adenosine on nucleotide synthesis from adenine has been studied in human erythrocytes.[57] Adenosine almost completely prevented the increase in total ATP that usually follows incubation with adenine in this system and inhibited incorporation of radioactive adenine into ATP by 94 percent. Quite high concentrations (12–16 m*M*) of adenosine had to be used, and it was concluded that adenosine itself, rather than a metabolite, was the inhibitory compound. By analogy with enzyme studies,[71] it was suggested that adenosine might be inhibiting adenine phosphoribosyltransferase.

If incubation with adenosine were to increase ATP concentrations in any system, inhibition of PP-ribose-P synthesis might be expected. However, because of the stimulation of PP-ribose-P synthesis in the presence of adenine that has been discussed above, it is not certain that this would inhibit ATP synthesis from adenine.

It is certain, however, that nucleotide synthesis from adenine and by adenosine and their analogues leads to the inhibiton of ATP synthesis *de novo*. This generally is called "feedback inhibition of purine biosynthesis *de novo*" and since 1946 has been demonstrated in a great many systems.[2] Although the name given to this phenomenon implies that a single mechanism is involved, and although thinking on this subject has tended to focus on end-product inhibition of amidophosphoribosyltransferase, three potential mechanisms need to be considered. In addition to the one just stated, these include end-product inhibition of PP-ribose-P synthetase and preferential utilization of PP-ribose-P by the exogenously supplied adenine.

There seems little question that the very process of nucleotide formation from adenine can lead to inhibition of *de novo* synthesis through competition for PP-ribose-P.[72,73] If this inhibitory factor is eliminated by removal of free adenine following elevation of ATP concentrations, however, the other two possibilities still remain. In studies using Ehrlich ascites tumor cells, it has been possible to study the effect of elevated and lowered ATP concentrations on the effective activity of both PP-ribose-P synthetase and amidophosphoribosyltransferase.[34,55] These experiments showed clearly that elevated ATP concentrations inhibited

PP-ribose-P synthetase and that lowered concentrations stimulated this activity. In contrast, neither change in ATP concentration affected the activity of amidophosphoribosyltransferase.

These conclusions, of course, need not necessarily apply to all cell types. It is of interest that a reduction of adenine nucleotide concentrations in rat liver *in vivo* was followed by a partial shift in the distribution of amidophosphoribosyltransferase subunits from the large inactive form to the small active form of the enzyme.[74] Although additional effects directly on PP-ribose-P synthetase were not ruled out, these findings are the first direct evidence that the nucleotide regulation of this enzyme that has been observed in enzyme studies may also apply to intact cells.

It remains to say, however, that at present there is more evidence from studies of intact cells for end-product inhibition of PP-ribose-P synthetase than for such inhibition of amidophosphoribosyltransferase. Technically, however, it is quite difficult to measure effects on the latter enzyme if there is also inhibition of the former.

Repression of Enzyme Synthesis

Although the pathway of purine biosynthesis *de novo* in microorganisms appears to be regulated by repression as well as by end-product inhibition,[2,75,76] there is evidence that repression is not a significant regulatory mechanism in mammalian cells.[77,78]

REFERENCES

1. K. O. Raivio and J. E. Seegmiller, *Curr. Top. Cell. Regul.* **2,** 201 (1970).
2. J. F. Henderson, *Regulation of Purine Biosynthesis,* American Chemical Society, Washington, DC (1972).
3. J. F. Henderson and A. R. P. Paterson, *Nucleotide Metabolism,* Academic, New York (1973).
4. J. B. Wyngaarden and W. N. Kelley, *Gout and Hyperuricemia,* Grune and Stratton, New York (1976).
5. E. W. Holmes, in *Uric Acid,* W. N. Kelley and I. M. Weiner, Eds., Springer, Berlin (1978), p. 21.
6. W. J. Arnold, in *Uric Acid,* W. N. Kelley and I. M. Weiner, Eds., Springer, Berlin (1978), p. 43.

7. J. F. Henderson, in *Uric Acid,* W. N. Kelley and I. M. Weiner, Eds., Springer, Berlin (1978), p. 75.
8. J. F. Henderson, in *Physiological and Regulatory Functions of Adenosine and Adenine Nucleotides,* H. P. Baer and G. I. Drummond, Eds., Raven, New York (1979), p. 315.
9. G. K. Smith, P. A. Benkovic, and S. J. Benkovic, *Biochemistry* **20,** 4034 (1981).
10. G. K. Smith, W. T. Mueller, P. A. Benkovic, and S. J. Benkovic, *Biochemistry* **20,** 1241 (1981).
11. J. F. Henderson, G. Zombor, and P. W. Burridge, *Can. J. Biochem.* **56,** 474 (1978).
12. P. Marko, E. Gerlach, H.-G. Zimmer, I. Pechan, T. Cremer, and C. Trendelenburg, *Hoppe-Segler's Z. Physiol. Chem.* **350,** 1669 (1969).
13. H.-G. Zimmer, C. Trendelenburg, and E. Gerlach, *J. Mol. Cell. Cardiol.* **4,** 279 (1972).
14. L. J. Fontenelle and J. F. Henderson, *Biochim. Biophys. Acta* **177,** 175 (1969).
15. P. C. L. Wong and J. F. Henderson, *Biochem. J.* **129,** 1085 (1972).
16. D. H. Namm, *Circ. Res.* **33,** 686 (1973).
17. C. M. Smith and J. F. Henderson, *Can. J. Biochem.* **54,** 341 (1976).
18. T. Hovi, A. C. Allison, and J. Allsop, *FEBS Lett.* **55,** 291 (1975).
19. K. O. Raivio and T. Hovi, *Exp. Cell Res.* **116,** 75 (1978).
20. F. F. Snyder, J. Mendelsohn, and J. E. Seegmiller, *J. Clin. Invest.* **58,** 654 (1976).
21. H. G. Zimmer and E. Gerlach, *Circ. Res.* **35,** 536 (1974).
22. H.-G. Zimmer and E. Gerlach, *Basis Res. Cardiol.* **72,** 241 (1977).
23. E. Rapaport and P. C. Zamecnik, *Proc. Natl. Acad. Sci.* (USA) **75,** 1145 (1978).
24. M. A. Becker, in *Uric Acid,* W. N. Kelly and I. M. Weiner, Eds. Springer, Berlin (1978), p. 155.
25. I. H. Fox and W. N. Kelley, *J. Biol. Chem.* **246,** 5739 (1971).
26. M. A. Becker, L. J. Meyer, W. H. Huisman, C. Lazar, and W. A. Adams, *J. Biol. Chem.* **252,** 3911 (1977).
27. L. J. Meyer and M. A. Becker, *J. Biol. Chem.* **252,** 3919 (1977).
28. E. W. Holmes, J. B. Wyngaarden, and W. N. Kelley, *J. Biol. Chem.* **248,** 6035 (1973).
29. W. W. Winder, R. J. Terjung, K. M. Baldwin, and J. O. Holloszy, *Am. J. Physiol.* **227,** 1411 (1974).
30. V. Schultz and J. M. Lowenstein, *J. Biol. Chem.* **251,** 485 (1976).
31. R. T. Bogusky, L. M. Lowenstein, and J. M. Lowenstein, *J. Clin. Invest.* **58,** 326 (1976).
32. M. B. Van der Weyden and W. N. Kelley, *J. Biol. Chem.* **249,** 7282 (1974).
33. F. F. Snyder and J. F. Henderson, *Can. J. Biochem.* **51,** 943 (1973).

34. J. Barankiewicz and J. F. Henderson, *Can. J. Biochem.* **55,** 257 (1977).
35. W. D. L. Musick, *CRC Crit. Rev. Biochem.* **11,** 1 (1981).
36. R. L. Miller, D. L. Adamczyk, and W. H. Miller, *J. Biol. Chem.* **254,** 2339 (1979).
37. C. M. Andres and I. H. Fox, *J. Biol. Chem.* **254,** 11388 (1979).
38. A. Y. Divekar and M. T. Hakala, *Mol. Pharmacol.* **7,** 663 (1971).
39. T. K. Leibach, G. I. Spiess, T. J. Neudecker, G. J. Peschke, G. Puchwein, and G. Hartmann, *Hoppe-Seylers Z. Physiol. Chem.* **352,** 328 (1971).
40. C. F. Hawkins, J. M. Kyd, and A. S. Bagnara, *Arch. Biochem. Biophys.* **202,** 380 (1980).
41. J. F. Henderson, A. Mikoshiba, S. Y. Chu & I. C. Caldwell, *J. Biol. Chem.* **247,** 1972 (1972).
42. I. H. Fox and W. N. Kelley, *Ann. Intern. Med.* **74,** 424 (1972).
43. H.-G. Zimmer and E. Gerlach, *Pflügers Arch.* **376,** 223 (1978).
44. JO Barankiewicz, M. L. Battell, and J. F. Henderson, *Can. J. Biochem.* **55,** 834 (1977).
45. J. Barankiewicz and J. F. Henderson, *Biochim. Biophys. Acta* **479,** 371 (1977).
46. S. D. Skaper, W. E. O'Brien, and I. A. Schafer, *Biochem. J.* **172,** 457 (1978).
47. P. Feigelson and M. Feigelson, *J. Biol. Chem.* **238,** 1073 (1963).
48. M. Feigelson, P. R. Gross, and P. Feigelson, *Biochem. Biophys. Acta* **55,** 495 (1962).
49. L. J. Fontenelle and J. F. Henderson, *Biochim. Biophys. Acta* **177,** 88 (1969).
50. G. W. Crabtree and J. F. Henderson, *Cancer Res.* **31,** 985 (1971).
51. J. K. Lowe, L. Brox, and J. F. Henderson, *Cancer Res.* **37,** 736 (1977).
52. C. E. Cass, J. K. Lowe, J. M. Manchak, and J. F. Henderson, *Cancer Res.* **37,** 3314 (1977).
53. F. F. Snyder and J. F. Henderson, *Cancer Res.* **33,** 2425 (1973).
54. F. F. Snyder and J. F. Henderson, *J. Cell. Physiol.* **82,** 349 (1973).
55. A. S. Bagnara, A. A. Letter, and J. F. Henderson, *Biochem. Biophys. Acta* **374,** 259 (1974).
56. E. M. Warrendorf and D. Rubenstein, *Blood* **42,** 637 (1973).
57. S. V. Manohar, M. H. Lerner, and D. Rubenstein, *Can. J. Biochem.* **46,** 445 (1968).
58. P. Lund, N. W. Cornell, and H. A. Krebs, *Biochem. J.* **152,** 593 (1975).
59. M. H. Lerner and D. Rubenstein, *Biochim. Biophys. Acta* **224,** 301 (1970).
60. W. Isselhard, J. Eitenmuller, W. Maurer, A. DeVreese, H. Reineke, A. Czerniak, J. Sturz, and H.-G. Herb, *J. Mol. Cell. Cardiol.* **12,** 619 (1980).
61. M. W. Seraydarian, L. Artaza, and B. C. Abbott, *J. Mol. Cell. Cardiol.* **4,** 477 (1972).
62. J. H. Fraser, H. Meyers, J. F. Henderson, L. W. Brox, and E. E. McCoy, *Clin. Biochem.* **8,** 353 (1975).

63. J. B. Wyngaarden and W. N. Kelley, in *Metabolic Basis of Inherited Disease*, 3rd ed., J. B. Stanbury, J. B. Wyngaarden, and D. S. Fredrickson, Eds., McGraw-Hill, New York (1972), p. 889.
64. E. Zoref-Shani and O. Sperling, *Biochim. Biophys. Acta* **607,** 503 (1980).
65. K. A. Ravio and J. E. Seegmiller, *Biochim. Biophys. Acta* **299,** 273 (1973).
66. M. S. Hershfield and J. E. Seegmiller, *J. Biol. Chem.* **251,** 7348 (1976).
67. A. S. Bagnara, L. W. Brox, and J. F. Henderson, *Biochim. Biophys. Acta* **350,** 171 (1974).
68. J. F. Henderson, J. K. Lowe, and J. Barankiewicz, in *Purine and Pyrimidine Metabolism,* Elsevier, Amsterdam (1977), p. 3.
69. W. L. Nyhan, L. Sweetman, and M. Lesch, *Metabolism* **17,** 846 (1968).
70. B. I. Uppen and P. G. Scholefield, *Can. J. Biochem.* **43,** 209 (1965).
71. M. Hori, R. E. A. Gadd, and J. F. Henderson, *Biochem. Biophys. Res. Commun.* **28,** 616 (1967).
72. J. F. Henderson and M. K. Y. Khoo, *J. Biol. Chem.* **240,** 2358 (1965).
73. J. F. Henderson and M. K. Y. Khoo, *J. Biol. Chem.* **240,** 3104 (1965).
74. M. Itakura, R. L. Sabina, P. W. Heald, and E. W. Holmes, *J. Clin. Invest.* **67,** 994 (1981).
75. R. K. Koduri and J. S. Gots, *J. Biol. Chem.* **255,** 9594 (1980).
76. R. A. Levine and M. W. Taylor, *Mol. Gen. Genet.* **181,** 313 (1981).
77. D. W. Martin, Jr. and N. T. Owen, *J. Biol. Chem.* 247, 5477 (1972).
78. M. W. Taylor, S. Olivelle, R. A. Levine, K. Coy, H. Hershey, K. C. Gupta, and L. Zawistowich, *J. Biol. Chem.* **257,** 377 (1981).

7

ATP Catabolism

PATHWAYS OF ATP CATABOLISM

The reactions of ATP catabolism are, in general, well worked out; however, their relative rates vary greatly in different biological systems, and their regulation in cells and tissues still is not well understood.[1,2] The alternative pathways of ATP catabolism involve four types of process: dephosphorylation; deamination; cleavage of the glycosidic bond; and oxidation and cleavage of the purine ring.

Dephosphorylation

The dephosphorylation of ATP may be divided into two stages, the conversion of ATP to adenylate and of adenylate and inosinate to adenosine and inosine, respectively:

$$\text{ATP} \rightarrow \text{ADP} \rightarrow \text{Adenylate} \tag{23}$$

$$\text{Adenylate} \rightarrow \text{Adenosine} + P_i \tag{24}$$

$$\text{Inosinate} \rightarrow \text{Inosine} + P_i \tag{25}$$

The first process, that of cleaving the high-energy phosphate bonds of ATP, may be catalyzed by reactions that utilize ATP, by adenylate kinase (EC 2.7.4.3), and by various specific and nonspecific phosphohydrolases; it is not always possible to identify exactly which enzymes are responsible for these processes in intact cells. The second process, that of cleaving the ribose–phosphate bond of the purine ribonucleoside monophosphates, also may be cleaved by several different enzymes. Thus a variety of so-called 5′-nucleotidases (EC 3.1.3.5) have been identified with different specificities and different cellular and subcellular distributions, and, in addition, various nonspecific phosphatases that can also catalyze this reaction are known to be widespread. The relative importance *in vivo* of all these enzymes simply is not well understood at the present time.

Deamination

This process occurs at two steps in the overall process of ATP catabolism:

$$\text{Adenylate} \rightarrow \text{Inosinate} + NH_4^+ \qquad (26)$$

$$\text{Adenosine} \rightarrow \text{Inosine} + NH_4^+ \qquad (27)$$

Both enzymes involved, adenylate deaminase (EC 3.5.4.6) and adenosine deaminase (EC 3.5.4.4), catalyze irreversible reactions and are highly specific with respect to their substrates. Their properties and regulation are discussed below.

In some plants and microorganisms, adenine itself may be deaminated to hypoxanthine.[3] However, the adenine deaminase (EC 3.5.4.2) reaction is not considered further here.

Cleavage of the Glycosidic Bond

Although the cleavage of inosine is the principal such reaction in ATP catabolism, that of adenosine must also be considered:

$$\text{Inosine} + P_i \rightleftharpoons \text{Hypoxanthine} + \text{Ribose-1-P} \qquad (28)$$

$$\text{Adenosine} + P_i \rightleftharpoons \text{Adenine} + \text{Ribose-1-P} \qquad (29)$$

Inosine cleavage, a readily reversible reaction, is catalyzed by purine nucleoside phosphorylase (EC 2.4.2.1). Guanosine is also a good substrate for this enzyme, and xanthosine is also cleaved, although at lower rates.

The rate and significance of adenosine cleavage in mammalian cells are not clear at the present time. This reaction can be catalyzed by purine nucleoside phosphorylase, although the Michaelis constant for adenine (the reaction often is assayed in the synthetic direction) is 10–20 times that for hypoxanthine.[4] However, evidence has also been presented that a specific adenosine phosphorylase may also exist in animal cells, with a relatively low Michaelis constant for adenine.[5] To complicate this picture further, the formation of adenine from adenosine by cultured cells has been proposed as a diagnostic test for mycoplasma infection.[6,7]

Technically, the cleavage of adenosine by intact cells is quite difficult to measure accurately, and deoxyadenosine sometimes is used as an alternative substrate;[8] however, it has not been proved that the same enzyme is involved in both cases. Certainly, the cleavage of deoxyadenosine has been demonstrated in mycoplasma-free cultured cells,[9,10] and under certain conditions it may be a major pathway of deoxyadenosine metabolism.[8–10]

Quantitatively, the cleavage of adenosine to adenine is a minor reaction; relative rates of the synthesis of adenosine, inosine, and guanosine from the respective purine bases plus ribose-1-P in extracts of rat liver were found to be 0.53, 139, and 190, respectively.[5]

In some plants and microorganisms, glycosidic bond cleavage takes place at the nucleotide level and is hydrolytic in nature; the products, for example, could be adenine plus ribose-5-P. This process is not considered further here.

Oxidation and Cleavage of the Purine Ring

If adenine is produced by cleavage of adenosine, it is not further catabolized in animal cells, although in intact animals it may be oxidized by xanthine oxidase to 2,8-dioxyadenine. Hypoxanthine produced from inosine, however, may be oxidized to uric acid, and the purine ring of the latter may then be cleaved:

$$\text{Hypoxanthine} \rightarrow \text{Xanthine} \rightarrow \text{Uric acid} \qquad (30)$$

$$\text{Uric acid} \rightarrow \text{Allantoin} \qquad (31)$$

Xanthine oxidase (EC 1.2.3.2) or xanthine dehydrogenase (EC 1.2.1.37) catalyzes the successive oxidation of hypoxanthine to xanthine and of xanthine to uric acid, whereas urate oxidase (EC 1.7.3.3) converts uric acid to allantoin. The latter enzyme is absent from humans and higher apes and from uricotelic animals but present in other mammals. It should be pointed out that xanthine can also be produced by the deamination of guanine, and hence it is an important end product of the catabolism of guanine nucleotides.

The overall scheme of ATP catabolism may be pictured as follows:

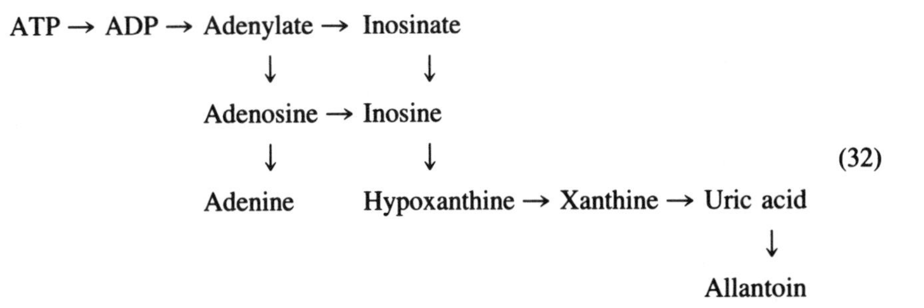

REGULATION OF ATP CATABOLISM AS A WHOLE

Three general questions may be raised concerning the overall process of ATP catabolism:

1. What different functions do the various pathwyas of ATP catabolism serve?
2. How is this process regulated so that ATP catabolism normally proceeds at only a low rate—that is, so that ATP is conserved and not wasted?
3. What influences or factors can initiate accelerated rates of ATP catabolism?

Functions of ATP Catabolic Pathways

Both ATP catabolism as a whole and the individual reactions and pathways of which it is composed may be regarded as having several distinct functions. Adenosine triphosphate catabolism, for example, plays a distinct role when it operates in the context of the preparation of foodstuffs and in their digestion and absorption; cells consumed in the diet contain ATP (or did originally), and this ATP is broken down to components that may be absorbed (see Chapter 9).

A second function is in the handling of dead cells within the body, such as skin and intestinal mucosa. Nuclei that are extruded and destroyed in the last stage of erythropoiesis may also contain ATP.

Adenosine triphosphate may also be considered to be "catabolized" by the many reactions in which it is converted to ADP and adenylate; however, these are better considered under the heading of "Metabolic Utilization of ATP Moieties" (see Chapter 5).

Finally, the catabolism of ATP as far as inosinate may have a biosynthetic function to the extent that the inosinate can be further converted to xanthylate and thence to guanine nucleotides.

Portions of the alternative pathways of ATP catabolism are also utilized for the catabolism of adenosine- or adenylate-containing cyclic nucleotides and coenzymes, of nucleotides produced by the hydrolysis of RNA, and of adenine-containing products of DNA repair. As with ATP itself, part of the catabolites are terminal end products whereas part are reutilized.

Finally, as mentioned in Chapter 7, the catabolism of inosinate is part of the main route of nitrogen excretion in uricotelic organisms.

Mechanism of ATP Conservation

It would seem that the principal mechanism for conserving ATP (i.e., for preventing catabolism) is the ability of cells effectively to carry out the phosphorylation of adenylate and ADP. Adenylate is the first substrate for effectively irreversible catabolic reactions, and its concentration normally is maintained at 1–5 percent (or less) that of ATP simply by what may be called the *phosphorylation potential* of cells. Any increase in its concentration or rate of synthesis can lead to its further metabolism by dephosphorylation or deamination. The role of allosteric regulation of the enzymes that carry out these catabolic reactions in the overall conservation of ATP is not certain at the present time.

The efficiency of the cellular processes that conserve ATP may be estimated from the rate of loss of radioactivity from ATP following labeling with radioactive adenine or adenosine. (Such data are difficult to interpret with precision because of various cyclic processes into which ATP enters; nevertheless, some estimates of net ATP turnover can be obtained.) In one study using rabbit erythrocytes, the turnover time (half-life) of ATP *in vivo* was estimated to be about 25 hours; this is in contrast to the half-life of the erythrocytes themselves, which was 10–12 days.[11]

The relative rates of catabolism and conservation of ATP may also be estimated by comparing the conversion of radioactive adenine to ATP and to inosine plus hypoxanthine during incubation of cells *in vitro* with this precursor. In an early study,[12] the ratio of ATP synthesis to ATP (actually adenylate) catabolism was 8–9. In isolated rate hepatocytes incubated with adenine, the ratio was approximately 3.[13]

Initiation of the Catabolic Process

Inasmuch as ATP normally is catabolized at relatively low rates, it seems appropriate to ask what factors, influences, or conditions act to accelerate this rate. In general, these seem to be of two types: (1) decreased ability of cells to phosphorylate adenylate and ADP and (2) the increased utilization of ATP for phosphorylation reactions: These, of course, are not unrelated.

It is clear from what has been said above that any condition that would alter the equilibrium among the three adenosine phosphates more in the direction of adenylate would favor the further catabolism of this nucleoside monophosphate, whether through deamination, dephosphorylation, or both. This shift in equilibrium may be due to (1) lack of substrates for substrate level or oxidative phosphorylation, lack of oxygen, or inhibition of some aspects of these processes or (2) decreased availability of P_i for the phosphorylation reactions. This, in turn, may be due to a generally low P_i concentration in the biological system under study or to any kind of trapping of phosphate derived from ATP or other phosphate donor.

To give just a few examples, ATP almost completely disappears when bovine erythrocytes are incubated in the absence of glucose,[14] and when Ehrlich ascites tumor cells are incubated anerobically without glucose.[15] Glycolytic inhibitors such as sodium fluoride[14] and iodoacetate[16] and inhibitors of oxidative phosphorylation such as dinitrophenol[17] and rotenone[18] also produce marked losses in ATP. Glycolysis can compensate for loss of oxidative phosphorylation, and the latter for the former, in maintaining ATP concentrations. Thus rotenone had no effect on ATP concentrations in the presence of glucose,[18] and cells incubated aerobically in the absence of glucose maintain high ATP levels if adequately oxygenated.[15]

The second type of detrimental influence is lowered P_i concentrations. At low intracellular concentrations, glycolysis is retarded, and hence so is the glycolytic regeneration of ATP.[19] Ehrlich ascites tumor cells incubated with 5 m*M* P_i, for example, have a lower ATP:ADP ratio than do those incubated with 15 or 25 m*M* P_i,[20] indicating that although some phosphorylation of ADP is occurring, this is not optimal. When such cells contain radioactive ATP, a slightly increased rate of catabolism to inosine and hypoxanthine can be detected.[21]

Another way to lower P_i concentrations is through "trapping" it in the form of some product of phosphorylation involving ATP. This in itself is simply one more instance of ATP utilization and would not ordinarily be regarded as initiating catabolism. However, it appears that catabolism can be greatly accelerated if the phosphorylated product is not metabolized and hence forms a permanent "sink" for P_i; alternatively, the phosphorylated product may be metabolizable, but only at slow rates. In addition, the formation of a well-metabolized phosphorylated compound may be detrimental if either the rate of phosphorylation is very rapid or the extent of phosphorylation is great; in these cases, there may be a transient but rapid or especially marked decrease in the ATP:ADP and

ATP:AMP ratios, leading to formation of adenylate, which is then further catabolized.

Deoxyglucose is an example of a compound that is readily phosphorylated by ATP but not metabolized; it thus permanently decreases the amount of P_i that is available for the phosphorylation of adenylate and ADP. Metabolizable substrates include fructose and glucose, which when presented to cells suddenly or at very high concentrations lead to rapid and extensive phosphorylation, transient decreases in ATP, and accelerated catabolism of adenylate (see Chapter 11 for references).

It should be apparent that in practice it is difficult to distinguish between initiation of ATP catabolism due to increased rates of phosphorylation and that due to an increased extent of phosphorylation, as in many situations both are involved. There is some evidence that each can individually be an initiating factor for catabolism, but this point requires further experimentation.

Regulation of Alternative Pathways of ATP Catabolism

Inasmuch as the adenylate formed during ATP catabolism may be metabolized by both deamination and by dephosphorylation, it is necessary to ask what the relative rates of these (and subsequent) alternative pathways are and how the several branch points are regulated.

Methodological Considerations

Before reporting experimental results that bear on these questions, it is necessary to consider several important methodological problems that complicate the assessment of the relative rates of deamination and dephosphorylation of adenylate.

Adenosine and inosinate formed from adenylate may both—at least potentially—be converted back to adenylate (and to ADP and ATP); adenosine kinase is involved in the former case and adenylosuccinate synthetase plus adenylosuccinate lyase, in the latter.

Leaving aside this question of resynthesis, it may be said that the relative net rates of deamination and dephosphorylation of adenylate may easily be assessed *if* the only metabolites that are formed are adenosine and inosinate. Further metabolism of adenosine to adenine or of inosinate to xanthylate and guanylate and their metabolites, if not actually measured, could, of course, lead to erroneous conclusions; in most cases, however, these are not major pathways.

A much more serious problem arises if inosine and its metabolites (hypoxanthine, xanthine, etc.) are formed as products of ATP catabolism, inasmuch as they can be formed from both inosinate and adenosine; analytical data alone seldom are sufficient to indicate the source of such inosine. This problem introduces considerable ambiguity into the assessment of the relative rates of adenylate deamination and dephosphorylation, and inasmuch as this was not always appreciated in early work on ATP catabolism, the conclusions of these studies sometimes need to be reevaluated.

One contemporary approach to this problem is to inhibit one route of inosine production by using one of several potent inhibitors of adenosine deaminase [e.g., coformycin, deoxycoformycin, erythro-9-(2-hydroxy-3-nonyl)adenine; see Chapter 11]. This approach has provided very valuable information concerning relative rates of alternative pathways of adenylate metabolism in several cell types, and some of these results are considered below.

Adenosine deaminase inhibitors should not be used uncritically, however, as at higher doses than those required to inhibit this enzyme, they may also inhibit other enzymes, including adenylate deaminase[22–26] and the conversion of inosinate to adenine and guanine nucleotides.[23] They may also compete with exogenously supplied adenosine for entry into cells.[23]

Relative Rates and Rate-Limiting Steps

Despite the methodological problems just discussed, it still is possible to discern considerable variation among cell and tissue types with respect to relative rates of alternative pathways of ATP catabolism; furthermore, for each tissue there may also be variation among species. Table 7.1 shows results of some older studies on ATP catabolism in which the original data have been normalized to show the relative extent of accumulation of adenylate, inosinate, and other catabolites (inosine, hypoxanthine, xanthine, and uric acid) as percentages of the total loss of ATP plus ADP. (These results are selected from much more extensive data just for the sake of illustration; the original papers should be consulted for further details.) Under the conditions used, it may seem that adenylate preferentially accumulates in some cases, inosinate in others, and inosine and its metabolites in still others; in some cases, two or more catabolites accumulate to roughly the same extent. The catabolites that accumulate in the highest concentrations may depend not only on the tissue, but also on the extent and duration of the catabolic process; time, therefore, may be an important variable. In addition, the pathway that is preferentially followed may depend on the conditions used to initiate catabolism; this point is elaborated on below.

Table 7.1. End Products of ATP Catabolism

		Metabolite Accumulated[a]				
Tissue	Species	Adenylate	Inosinate	Adenosine	Inosine + Hypoxanthine	Reference
Brain	Rat	86	5	9	8	27
	Rabbit	55	2	17	36	28
Kidney	Rat	51	19	—	5	29
	Dog	43	27	—	28	30
Liver	Dog	64	4	3	37	30
	Rat	78	2	—	12	27
Skeletal muscle	Rat	2	66	—	27	27
	Rabbit	3	82	1	14	28
Heart	Rat	48	2	13	27	27
	Guinea pig	32	3	2	63	29
	Cat	9	2	1	88	29

[a]Percent of ATP plus ATP catalyzed.

With the availability of inhibitors of adenosine deaminase, adenosine kinase, and in a few cases adenylate deaminase as well, it has more recently become possible to study the alternative pathways of ATP catabolism in greater detail and to resolve some of the ambiguities inherent in earlier studies.

The first such study was that by Lomax and Henderson,[31] who used Ehrlich ascites tumor cells incubated *in vitro* with deoxyglucose to accelerate the rate of ATP catabolism. By measuring catabolite concentrations in the presence and absence of an adenosine deaminase inhibitor, it was possible to calculate that at most 18 percent of the net adenylate that was metabolized was dephosphorylated. When ATP catabolism was accelerated by use of deoxyglucose, the inosinate formed by the deamination of adenylate was mostly dephosphorylated to inosine. In addition, a small amount of cycling between adenylate and adenosine was detected, amounting to approximately 10 percent of the total amount of ATP catabolized. Further studies using these tumor cells,[17] however, showed that a different inducer of ATP catabolism (dinitrophenol in the absence of glucose) produced a different pattern of catabolism; in this case, there was very little dephosphorylation of either inosinate or adenylate.

Deoxyglucose and inhibitors of oxidative phosphorylation have also been compared in cultured human lymphoblasts.[18] When deoxyglucose was used to accelerate ATP catabolism, about 85 percent of adenylate metabolism was by means of deamination, but when ATP catabolism was induced by anaerobosis or rotenone, 60–80 percent was metabolized by dephosphorylation.

Different patterns of ATP catabolism have been observed by use of erythrocytes. When human erythrocytes were incubated with deoxyglucose and an adenosine deaminase inhibitor, about 65 percent of the net adenylate formed simply accumulated; of the amount metabolized, however, approximately 30 percent was deaminated and 70 percent dephosphorylated. However, when the purine nucleoside analogue tubercidin was used to accelerate ATP synthesis, little adenylate accumulated and virtually all the adenylate was metabolized by deamination to inosinate; although some of this inosinate accumulated, most was dephosphorylated.[16] In another study of human erythrocytes (in this case, cells genetically deficient in adenosine deaminase), ATP catabolism induced by fluoride predominantly followed the pathway: adenylate → inosinate → inosine → hypoxanthine. However, when iodoacetate was used, most of the adenylate formed was dephosphorylated.[32]

Finally, in liver cells both *in vivo*[26] and *in vitro*,[33] the inhibition of adenosine deaminase had no effect on the catabolism of ATP induced by fructose, and there was no accumulation of adenosine; in this case, therefore, no dephosphorylation of adenylate occurred. It is of interest that if concentrations of the inhibitor were used that inhibited both adenosine deaminase and adenylate deaminase,[26] there was no net accumulation of adenylate. Thus, even though adenylate must have been formed as a result of the massive phosphorylation of fructose, whatever adenylate and ADP was produced was rephosphorylated in the absence of an available catabolic route.

These results amply demonstrate that the relative rates of the alternative pathways of ATP catabolism vary greatly; furthermore, these vary not only with cell type and species, but also with respect to the means used to accelerate the catabolic process. Under conditions of good nutrition and ample oxygenation, the first and principal rate-limiting step in ATP catabolism is the formation of adenylate; here the ordinary obstacle to catabolism is the ability of cells to keep the adenine nucleotide pool at a high degree of polyphosphorylation. Once catabolism proceeds to adenylate and beyond, however, the reaction or reactions that are rate limiting vary with both the biological system and the experimental conditions employed.

Individual Enzymes of Catabolism

Although it is not clearly understood what factors regulate the alternative pathways of ATP catabolism in any of the examples just presented, those that may at least potentially be of importance include the properties of the individual enzymes, including enzyme amount, subcellular localization, Michaelis constants in relation to substrate concentrations, and regulatory effects of end products and other metabolites.

Adenylate Deaminase

This enzyme has been very extensively studied[3,34] and is believed to be important in the regulation of ATP catabolism. The activity of this enzyme varies widely among tissues and in the rat skeletal muscle has much higher activity than other tissues;[3,35] activity is low in liver and in heart. Changes in total activity, physical properties, and isozyme pattern of adenylate deaminase during embryonic and postnatal development have been noted,[35–38] as have changes due to exercise[39] and diet.[40]

Adenosine triphosphate and GTP can both stimulate and inhibit adenylate deaminase activity, depending on their concentrations and other conditions.[41,42] In studies in which ATP and GTP concentrations in intact cells were raised[43] and lowered,[44] adenylate deaminase activity (measured in the intact cells) was slightly decreased when GTP was elevated but remained unchanged under all other conditions. However, the activity of this enzyme is low and difficult to measure accurately under these conditions.

5′-Nucleotidase

This single enzyme name covers a diverse group of enzymes, which differ in physical and kinetic properties and in cellular and subcellular localization.[1,2,45] Some prefer adenylate as substrate, whereas inosinate or guanylate are better substrates for others; some forms but not others are inhibited by nucleosides, nucleotides, or P_i. Many studies have dealt with ecto or at least membrane-bound 5′-nucleotidases, whereas at the present time effects are being made to identify and characterize the intracellular, soluble forms of this enzyme family.

Adenosine Deaminase

The activity of this enzyme varies widely among tissues and among species.[3,46,47] Although these variations may play a role in determining the rate of metabolism of exogenously supplied adenosine (and the relative rate of deamination and phosphorylation), there is no evidence that this enzyme has an important role in the regulation of ATP catabolism. Inosine acts as a product inhibitor of adenosine deaminase, but with K_i values of 0.6–1.6 mM.[48]

Purine Nucleoside Phosphorylase

This enzyme also is unlikely to play an important role in the regulation of the overall process of ATP catabolism.[49]

Relative Activities

As would be expected, the relative activities of the various catabolic enzymes vary among tissues, just as do the total activities of the individual enzymes. Table 7.2 reports some values for four enzymes of ATP catabolism and two enzymes that provide alternative routes of metabolism (see next section) from a study in which they were all measured simultaneously.[50] Although the ratio of nucleotidase activity on adenylate and inosinate were the same in liver and fibroblasts, the relationships between adenosine deaminase activity and those of the nucleotidases and between purine nucleoside phosphorylase and adenosine deaminase were quite distinct in each tissue. Similarly, the balance between the total activity of catabolic enzymes and the relevant alternative anabolic reaction varies widely.

Alternative Metabolic Pathways

In addition to the enzymes of ATP catabolism themselves, the substrates of these enzymes may be metabolized by other routes as well. These alternative pathways can be quite important in the regulation of ATP catabolism inasmuch as they

Table 7.2. Tissue Distribution of Catabolic and Anabolic Enzymes[a]

	Liver	Erythrocytes	Fibroblasts
Catabolic enzymes			
Adenylate nucleotidase	225	0	2340
Inosinate nucleotidase	190	0	1984
Adenosine deaminase	588	23	728
Purine nucleoside phosphorylase	1460	330	422
Alternative anabolic enzymes			
Adenosine kinase	2706	12	69
Hypoxanthine phosphoribosyltransferase	865	81	38

[a]From Shenoy and Clifford.[50]

alter the concentrations and availability of the intermediates of the catabolic pathway.

Adenylate

In addition to the catabolic pathways of deamination and dephosphorylation, adenylate can also, of course, be metabolized by phosphorylation. It has already been mentioned, for example, that in liver cells in which adenylate deaminase is inhibited, fructose treatment leads to no net change in the relative concentrations of ATP, ADP, and adenylate; this must be due in large part to the rephosphorylation of adenylate and ADP. In addition, phosphorylation can compete with deamination even when the latter pathway is operative. Thus in tumor cells treated with deoxyglucose to accelerate ATP catabolism, inhibition of adenosine deaminase to promote the resynthesis of adenylate from adenosine did not increase the deamination of adenylate, but rather its phosphorylation.[31]

Adenosine

The phosphorylation of adenosine by adenosine kinase (EC 2.7.1.20) provides as alternative to the catabolic process of deamination. The significance of this anabolic pathway, even when ATP catabolism is greatly accelerated, has been shown by the use of adenosine kinase inhibitors. In tumor cells treated with deoxyglucose to initiate ATP catabolism, ATP concentrations were appreciably lower in the presence of such an inhibitor than in its absence.[17]

The relative rates of deamination and phosphorylation of exogenously supplied adenosine have been studied in some detail.[51] The Michaelis constants for adenosine deaminase generally are lower than those for adenosine kinase, and on this basis low concentrations of adenosine might be expected to be preferentially phosphorylated; however, values for both constants vary appreciably from cell to cell. Two common patterns of adenosine metabolism have been distinguished on the basis of the relative rates of deamination and phosphorylation as a function of adenosine concentration. In one, deamination predominates at all adenosine concentrations; this has been observed in unstimulated human peripheral lymphocytes and in sheep erythrocytes. In the second, phosphorylation predominates at low adenosine concentrations whereas deamination is favored at high substrate concentrations; this occurs in Ehrlich ascites tumor cells, phytohemagglutinin-stimulated human lymphocytes, and human erythrocytes. In addition, mutant cells have been studied that lack either adenosine kinase or adenosine deaminase.

Inosinate

In addition to dephosphorylation, inosinate can be metabolized to both adenylosuccinate and adenylate and to xanthylate and guanylate. The former pathway is discussed elsewhere as a central portion of the "purine nucleotide cycle" (Chapter 9) and is not considered further here. The oxidation of inosinate to xanthylate and its further conversion to guanine nucleotides was readily demonstrable in tumor cells incubated with deoxyglucose;[17] however, this pathway constituted only about 10 percent of the total metabolsim of inosinate, as most was dephosphorylated. When a similar extent of ATP was generated under conditions in which the dephosphorylation of inosinate was substantially blocked (by the use of dinitrophenol), the rate of oxidation of inosinate increased and was approximately 45 percent of the total metabolism of inosinate; in addition, inosinate accumulated under these conditions.[17]

In some cells, inosinate is also converted to inosine triphosphate.[52] This pathway is generally believed to be a minor route of inosinate metabolism, but its rate has not really been accurately measured.

Inosine

Catabolism is the only fate of inosine in most mammalian cells, inasmuch as the adenosine deaminase reaction is irreversible, and phosphorylation either does not occur at all or only at a low rate. In some insects, however, the rate of phosphorolysis is low and inosine kinase (EC 2.7.1.73) appears to be present; here the phosphorylation of inosine appears to be a quite important synthetic reaction.[53]

Hypoxanthine

This purine base may potentially be catabolized through oxidation to xanthine and uric acid, converted to inosinate by hypoxanthine phosphoribosyltransferase (EC 2.4.2.8), and converted to inosine by purine nucleoside phosphorylase. In systems in which inosine cannot be phosphorylated, the latter route would simply be a dead end or constitute part of a futile cycle. Furthermore, the purine nucleoside phosphorylase reaction does not appear to operate in the synthetic direction in intact cells.[54]

It has been amply demonstrated that treatment of whole animals or human patients with inhibitors of xanthine oxidase very considerably increases the utilization of hypoxanthine for nucleotide formation[55]; this is also true of isolated hepatocytes *in vitro*.[56] That this hypoxanthine can be used for the synthesis of nucleic acid adenine[57] indicates that it can be converted, at least in part, to ATP. It must be recalled that the tissue distribution of xanthine oxidase is quite narrow; most of it is in the liver, with smaller amounts in intestine and kidney.[58,59] Thus in many tissues and cultured cells, treatment with xanthine oxidase inhibitors would be expected to have no effect on the metabolism and hypoxanthine. This conclusion is vitiated, however, by consistent observations that allopurinol can be converted to its ribonucleoside–5′-phosphate derivative by hypoxanthine phosphoribosyltransferase. This process itself consumes PP-ribose-P, and the allopurinol nucleotide formed can inhibit PP-ribose-P synthetase; as a consequence, PP-ribose-P concentrations are lowered.[55]

Adenosine triphosphate concentrations in liver and kidney of rats treated with allopurinol show only transient changes; in the liver, there is a slight decrease at one hour with a return to normal at three hours. In the kidney, ATP concentrations are slightly elevated at one hour and return to normal at three hours.[60]

Regulatory Mechanisms in Intact Cells

Two questions may be asked concerning the regulation of the alternative pathways of ATP catabolism in intact cells: (1) to what extent do potential regulatory factors identified in studies of cell-free preparations apply to regulation in intact cells and (2) whether any additional factors are important for regulating the rates of these reactions in intact cells.

Application of Enzyme Studies to Intact Cells

The fact that such very diverse patterns of ATP catabolism have been observed in studies using different cell types and varied experimental conditions makes it clear that the regulation of this process is complex and suggests, furthermore, that the relative importance of different regulatory mechanisms may vary. It may, therefore, be difficult if not impossible to deduce or propose a unified scheme for the regulation of ATP catabolism that will fit all cases.

Qualitatively, at least, some observations using intact cells are compatible with enzyme studies that show that soluble 5′-nucleotidase is inhibited by ATP[61] and that adenylate deaminase can be both activated and inhibited by GTP and ATP, depending on conditions, and inhibited by P_i.[62–65] However, these regulatory mechanisms cannot account for all whole-cell observations, and there still remain a number of inconsistencies and incongruities.

In one study that attempted quantitatively to test the application of enzyme studies to ATP catabolism in intact cells, Lomax et al.[17] concluded that ATP concentrations, for example, did not appear to affect adenylate deaminase activity. Thus for this enzyme from tumor cells, activation by ATP was maximal at concentrations of 300–500 μM, and higher concentrations inhibited deaminase activity; 50 percent inhibition was achieved at 900 μM.[62] Concentrations of ATP normally present in cells are definitely in the inhibitory range, and only after six minutes of incubation with 5.5 mM deoxyglucose do total ATP concentrations decrease to 500 μM; this is after a large fraction of the ATP originally present has already been converted to adenylate and deaminated. It thus seems unlikely that decreased inhibition of adenylate deaminase by ATP is an important regulatory factor under these conditions.

Similarly, the K_i for inhibition of tumor cell 5′-nucleotidase by ATP is 230 μM,[62] whereas the lowest ATP achieved after deoxyglucose treatment is about 400 μM.[17] Therefore, release of inhibition of 5′-nucleotidase activity by ATP is unlikely to play an important regulatory role in the dephosphorylation of inosinate during ATP catabolism in these cells.

These studies remain incomplete in that P_i concentrations were not measured, and it really is not certain which enzyme(s) is (are) responsible for the dephosphorylation of inosinate. However, these studies show the complexity of the regulation of ATP catabolism and raise doubts about conclusions based on qualitative changes alone; they also provide an approach to quantitative studies of this problem.

Regulatory Mechanisms of Intact Cells

Intact cells, of course, differ from the usual *in vitro* enzyme assay system in several respects, including enzyme concentration, presence of alternative pathways of both synthesis and utilization of substrates, and loss of compounds from cells into the external medium.

The high concentrations of enzyme that are present in intact cells means that the amount of substrate metabolized per unit time is very much greater than in enzyme assay systems. In a static system, an initially moderate substrate concentration may quickly be reduced to levels substantially below the Michaelis constant, and the rate of the reaction would rapidly decrease. In the cell, however, substrate is continually being provided by distal synthetic reactions, and this rate of provision of substrate may become as important—or more important—for the overall rate of reactions as the actual concentration of the substrate.

This idea was given experimental support in the studies by Lomax et al., mentioned earlier.[17] Here it was found that the rate of deamination of adenylate was related to the rate of synthesis of adenylate, but not to the actual concentration of this nucleotide. This was true also of the dephosphorylation of inosinate, the dephosphorylation of adenylate, and the dehydrogenation of inosinate. In these three cases, however, very high rates of substrate formation appeared to exceed the capacity of the enzyme to process them. The relative importance of substrate concentration and rate of substrate synthesis for the overall rate of various aspects of ATP catabolism needs to be tested in other biological systems.

Finally, the loss of purines from cells needs to be considered, as this is a very important determinant of the concentration of some metabolites. For cells incubated or grown *in vitro*, this loss is into the medium, which in most systems has a vastly greater volume than that of the intracellular space. In animals, loss is into extracellular space and blood. Although their combined volume is not so great proportionately as that of tissue water, circulation and excretion help provide very important routes of loss.

Inasmuch as purine nucleotides generally do not leak out of cells (at least at appreciable rates), the loss under consideration here applies to purine nucleosides and bases. Because the concentrations of such intermediates are often measured in intracellular and extracellular water together, the importance of loss from cells has seldom been considered.

Catabolism of Extracellular Adenine Nucleotides

Although ATP is usually regarded as an intracellular metabolite—and most of it is in this form—some ATP, ADP, and adenylate may be produced extracellularly as part of physiological processes such as platelet aggregation, release of neurotransmitters and neurohormones, and so on (see Chapter 8). These nucleotides are very rapidly catabolized in the extracellular space, and these reactions are considered briefly here.

The most important aspect of the catabolism of extracellular adenine nucleotides is their dephosphorylation to adenosine by plasma membrane enzymes whose active sites face the external medium rather than the cell cytoplasm ("ecto enzymes"). Phosphatases hydrolyzing ATP, ADP, and adenylate have been shown to exist as ecto enzymes in one or another biological system; in addition, adenylate deaminase and phosphatases acting on inosinate may exist on the cell surface.[66] The exact distribution of these enzymes and their relative rates, of course, vary among cell and tissue types studied. For example, cultured human endothelial cells catabolize ATP and ADP more rapidly than does adenylate, whereas just the opposite is true of cultured human fibroblasts.[67]

REFERENCES

1. I. H. Fox, in *Uric Acid,* W. N. Kelley and I. M. Weiner, Eds. Springer, Berlin (1978), p. 93.
2. I. H. Fox, *Metabolism* **30,** 616 (1981).
3. C. L. Zielke and C. H. Suelter, in *The Enzymes,* 3rd ed., vol. 4, P. D. Boyer, Ed. Academic, New York (1971), p. 47.
4. T. P. Zimmerman, N. B. Gersten, A. F. Ross, and R. P. Miech, *Can. J. Biochem.* **49,** 1050 (1971).
5. A. Y. Divekar, *Biochim. Biophys. Acta* **422,** 15 (1976).
6. M. A. Thomas, C. Shipman, Jr., J. N. Sandberg, and J. C. Drach, *In Vitro* **13,** 502 (1977).
7. M. Hatonaka, R. Del Giudice, and C. Long, *Proc. Natl. Acad. Sci. (U.S.A.)* **72,** 1401 (1975).
8. F. F. Snyder and J. F. Henderson, *J. Biol. Chem.* **248,** 5899 (1973).
9. D. Hunting, J. Hordern, and J. F. Henderson, *Can. J. Biochem.* **59,** 838 (1981).
10. J. Hordern and J. F. Henderson, *Can. J. Biochem.* (in press).
11. J. Mager, A. Hershko, R. Zeitlin-Beck, T. Shoshoni, and A. Razin, *Biochim. Biophys. Acta* **149,** 50 (1967).
12. G. W. Crabtree and J. F. Henderson, *Can. J. Biochem.* **49,** 959 (1971).
13. C. M. Smith, L. M. Rovamo, M. P. Kekomaki, and K. O. Raivio, *Can. J. Biochem.* **55,** 1134 (1977).
14. R. C. Smith, *Comp. Biochem. Physiol.* **69B,** 505 (1981).
15. J. F. Henderson, M. L. Battell, G. Zombor, and M. K. Y. Khoo, *Cancer Res.* **37,** 3434 (1977).
16. J. F. Henderson, G. Zombor, P. W. Burridge, G. Barankiewicz, and C. M. Smith, *Can. J. Biochem.* **57,** 873 (1979).
17. C. A. Lomax, A. S. Bagnara, and J. F. Henderson, *Can. J. Biochem.* **53,** 231 (1975).
18. S. S. Matsumoto, K. O. Raivio, and J. E. Seegmiller, *J. Biol. Chem.* **254,** 8956 (1979).
19. R. Wu and E. Racker, *J. Biol. Chem.* **234,** 1029 (1959).
20. E. H. Creaser, R. P. de Leon, and P. G. Scholefield, *Cancer Res.* **19,** 705 (1959).

21. J. F. Henderson and K. L. Boldt, unpublished results (1980).
22. R. P. Agarwal and R. E. Parks, Jr., *Biochem. Pharmacol.* **26,** 663 (1977).
23. J. F. Henderson, L. Brox, G. Zombor, D. Hunting, and C. A. Lomax, *Biochem. J.*, **188,** 913 (1980).
24. G. Van den Berghe, F. Bontemps, and H.-G. Hers, *Biochem. J.* **188,** 913 (1980).
25. M. Debatisse, M. Berry, and G. Buttin, *J. Cell. Physiol.* **106,** 1 (1981).
26. D. Siepen and J. F. Henderson, unpublished results (1981).
27. B. Deuticke, E. Gerlach, and R. Dierkesmann, *Pflügers Arch.* **292,** 239 (1966).
28. R. F. von Matthias and E.-W. Busch, *Hoppe-Seylers Z. Physiol. Chem.* **350,** 1410 (1969).
29. E. Gerlach, B. Deuticke, R. H. Dreissbach, and C. W. Rosarius, *Pflügers Arch.* 278, 296 (1963).
30. E.-W. Busch, I. M. von Dorcke, and B. Martinez, *Biochim. Biophys. Acta* **166,** 547 (1968).
31. C. A. Lomax and J. F. Henderson, *Cancer Res.* **33,** 2825 (1973).
32. G. C. Mills, R. M. Goldblum, and F. C. Schmalstieg, *Life Sci.* **29,** 1811 (1981).
33. C. M. Smith, L. M. Rovamo, and K. O. Ravio, *Can. J. Biochem.* **55,** 1237 (1977).
34. J. F. Henderson, in *Uric Acid,* W. N. Kelley and I. M. Weiner, Eds., Springer, Berlin (1978), p. 75.
35. J. M. Lowenstein, *Physiol. Rev.* **52,** 382 (1972).
36. H. Kluge and V. Wieczorek, *Acta Biol. Med. Germ.* **21,** 271 (1968).
37. N. Ogasawara, H. Gato, and T. Watanabe, *FEBS Lett.* **58,** 245 (1975).
38. K. Kalisha, *Internatl. J. Biochem.* **6,** 471 (1975).
39. W. W. Winder, R. L. Terjung, K. M. Baldwin, and J. O. Halloszy, *Am. J. Physiol.* **227,** 1411 (1974).
40. L. V. Turner and E. B. Fern, *Br. J. Nutr.* **32,** 539 (1974).
41. J. S. Gots, *Metabolic Pathways,* 3rd ed., vol. 5, H. J. Vogel, Ed., Academic, New York (1971), p. 225.
42. B. D. Sanwal, M. Kapoor, and H. W. Duckworth, *Curr. Top. Cell. Regul.* **3,** 1 (1971).
43. F. F. Snyder and J. F. Henderson, *Can. J. Biochem.* **51,** 943 (1973).
44. J. Barankiewicz and J. F. Henderson, *Can. J. Biochem.* **55,** 257 (1977).
45. G. I. Drummond and M. Yamamoto, in *The Enzymes,* 3rd ed., vol. 4, P. D .Boyer, Ed., Academic, New York (1971), p. 337.
46. P. E. Daddona and W. N. Kelley, *Mol. Cell. Biochem.* **29,** 91 (1980).
47. K. J. Collier, Jr. and P. F. Ma, *Internatl. J. Biochem.* **12,** 659 (1980).
48. M. B. Van der Weyden and W. N. Kelley, *J. Biol. Chem.* **251,** 5448 (1976).
49. R. E. Parks, Jr. and R. P. Agarwal, *The Enzymes,* 3rd ed., vol. 7, P. D. Boyer, Ed., Academic, New York (1971), p. 483.

50. T. S. Shenoy and A. J. Clifford, *Biochim. Biophys. Acta* **411,** 133 (1975).
51. J. F. Henderson, in *Physiological and Regulatory Functions of Adenosine and Adenine Nucleotides,* H. P. Baer and G. I. Drummond, Eds., Raven, New York (1979), p. 315.
52. J. H. Fraser, H. Meyers, J. F. Henderson, L. W. Brox, and E. E. McCoy, *Clin. Biochem.* **8,** 353 (1975).
53. M. M. Johnson, D. Nash, and J. F. Henderson, *Comp. Biochem. Physiol.* **66B,** 555 (1980).
54. J. Barankiewicz and J. F. Henderson, *Biochim. Biophys. Acta* **479,** 371 (1977).
55. G. B. Elion, in *Uric Acid,* W. N. Kelley and I. M. Weiner, Eds., Springer, Berlin (1978), p. 485.
56. M. Lalanne and F. Lafleur, *Can. J. Biochem.* **58,** 607 (1980).
57. R. Pomales, S. Bieber, R. Friedman, and G. H. Hitchings, *Biochim. Biophys. Acta* **72,** 119 (1963).
58. U. A. S. Al-Khalidi and T. H. Chaglassian, *Biochem. J.* **97,** 318 (1965).
59. R. W. E. Watts, J. E. M. Watts, and J. E. Seegmiller, *J. Lab. Clin. Med.* **66,** 688 (1965).
60. G. B. Elion and D. J. Nelson, *Adv. Exp. Biol. Med.* **41B,** 639 (1974).
61. A. W. Murray and B. Friedricks, *Biochem. J.* **111,** 83 (1969).
62. M. R. Atkinson and A. W. Murray, *Biochem. J.* **104,** 10C (1967).
63. T. J. Wheeler and J. M. Lowenstein, *J. Biol. Chem.* **254,** 8994 (1979).
64. J. G. Brady and J. F. Costello, *Biochim. Biophys. Acta* **350,** 455 (1974).
65. A. G. Chapman, A. L. Miller, and D. E. Atkinson, *Cancer Res.* **36,** 1144 (1976).
66. J. F. Manery and E. E. Dryden, in *Physiological and Regulatory Functions of Adenosine and Adenine Nucleotides,* H. P. Baer and G. I. Drummond, Eds., Raven, New York (1979), p. 323.
67. A. M. Dosne, C. Legrand, B. Bauvois, E. Bodevin, and J. P. Caen, *Biochem. Biophys. Res. Commun.* **85,** 183 (1978).

8

Metabolism of the Adenosine Moiety of ATP in Intact Cells and Animals

A number of important features of the metabolism of the adenosine moiety of ATP do not fit neatly into the categories of "ATP synthesis" and "ATP catabolism," nor are they properly pathological or pharmacological effects on ATP metabolism. It seems appropriate here, therefore, to outline (1) some basic functions of ATP and the utilization of its adenosine moiety at the cellular level, (2) some factors that affect ATP metabolism in tissues and animals, (3) some specialized functions of ATP that seem to be exhibited more at the level of tissues rather than individual cells, (4) some cyclic pathways of ATP utilization and synthesis, and (5) some aspects of the relationship between ATP catabolism and ATP synthesis in tissues.

Most of these topics are considered only briefly; space does not permit a thorough discussion of each. However, a few seem especially important and rather closely related to topics covered in other chapters, and these are given more weight.

BASIC FUNCTIONS OF ATP AND ITS UTILIZATION AT THE CELLULAR LEVEL

The adenosine moiety of ATP is used in a variety of biosynthetic and regulatory processes that can be listed only briefly.

RNA Synthesis

Adenosine triphosphate is incorporated into RNA together with GTP, UTP, and CTP by several RNA polymerases; each of the several species of RNA of course has its own rate of synthesis and of turnover. In addition, some RNA species terminate in sizable polyadenylate stretches. The actual amount of ATP that is represented by the adenylate moiety of RNA considerably exceeds the amount of free ATP in cells.

DNA Synthesis

For incorporation into DNA, the ribose of ATP must be converted to 2′-deoxyribose. This process is catalyzed by the enzyme ribonucleotide reductase, which uses ADP as substrate and forms deoxy-ADP as product. This is then converted to deoxy-ATP, which can then be incorporated into DNA. As with RNA, the amount of the (deoxy)adenylate moiety of DNA exceeds the amount both of ATP and deoxy-ATP in cells. In addition, concentrations of ATP are in excess of those of deoxy-ATP by 500–1000-fold in many cells.

The metabolism of deoxy-ATP can be outlined briefly as follows:

$$
\begin{array}{ccc}
\text{ATP} \rightarrow \text{ADP} \rightarrow \text{Deoxy-ADP} \rightarrow & \text{Deoxy-ATP} & \\
 & \updownarrows & \\
 & \text{Deoxy-AMP} & \\
 & \updownarrows & \\
 & \text{Deoxyadenosine} \rightarrow & \text{Deoxyinosine} \\
 & \downarrow & \downarrow \\
 & \text{Adenine} & \text{Hypoxanthine}
\end{array}
\qquad (33)
$$

Synthesis of Guanine Nucleotides

In cells in which all the various alternative pathways of purine nucleotide synthesis are operating, only small amounts of adenine nucleotides are converted to guanine nucleotides. However, when other pathways are blocked, cells can synthesize all their guanine nucleotides from adenylate. The pathway is as follows:

$$\begin{array}{l} \text{ATP} \rightarrow \text{ADP} \rightarrow \text{Adenylate} \\ \qquad\qquad\qquad\qquad\qquad \downarrow \\ \qquad\qquad\qquad \text{Inosinate} \rightarrow \text{Xanthylate} \rightarrow \text{Guanylate} \end{array} \tag{34}$$

The inosinate dehydrogenase reaction requires NADP, and the guanylate synthetase (sometimes also called *xanthylate aminase*) reaction requires ATP plus glutamine.

Synthesis of Coenzymes

The adenosine moiety of ATP is used in the synthesis of many essential coenzymes. The following are some of these reactions:

$$\text{ATP} \rightarrow \text{3}',\text{5}'\text{-Cyclic adenylate} + \text{PP}_i \tag{35}$$

$$\text{Nicotinamide ribonucleotide} + \text{ATP} \rightarrow \text{Nicotinamide adenine dinucleotide} + \text{PP}_i \tag{36}$$

$$\text{Nicotinate ribonucleotide} + \text{ATP} \rightarrow \text{Nicotinate adenine dinucleotide} + \text{PP}_i \tag{37}$$

$$\text{NAD} + \text{ATP} \rightarrow \text{NADP} + \text{ADP} \tag{38}$$

$$\text{Flavin mononucleotide} + \text{ATP} \rightarrow \text{Flavin adenine dinucleotide} + \text{PP}_i \tag{39}$$

$$\text{ATP} + \text{Methionine} \rightarrow \text{S-adenosylmethionine} + \text{PP}_i + \text{P}_i \tag{40}$$

$$\text{ATP} + 4'\text{-Phosphopanthetheine} \rightarrow \text{Dephospho CoA} + \text{PP}_i \qquad (41)$$

$$\text{Dephospho coenzyme A} + \text{ATP} \rightarrow \text{CoA} + \text{ATP} \qquad (42)$$

$$\text{ATP} + \text{SO}_4^{2-} \rightarrow \text{Adenosine phosphosulfate} + \text{PP}_i \qquad (43)$$

$$\text{Adenosine phosphosulfate} + \text{ATP} \rightarrow \text{Phosphoadenosine phosphosulfate} + \text{ATP} \qquad (44)$$

$$\text{Cobalamin} + \text{ATP} \rightarrow 5'\text{-Deoxyadenosyl} \qquad (45)$$

Poly ADP Ribose

The ADP–ribose portion of NAD can be transferred to proteins or polymerized to poly(ADP–ribose) stretches that are linked to proteins. Adenosine diphosphate–ribose itself is simply formed from NAD by hydrolysis; this moiety may also be transferred to an appropriate acceptor instead of to water. Finally, poly(ADP–ribose) can be formed with the following type of structure:

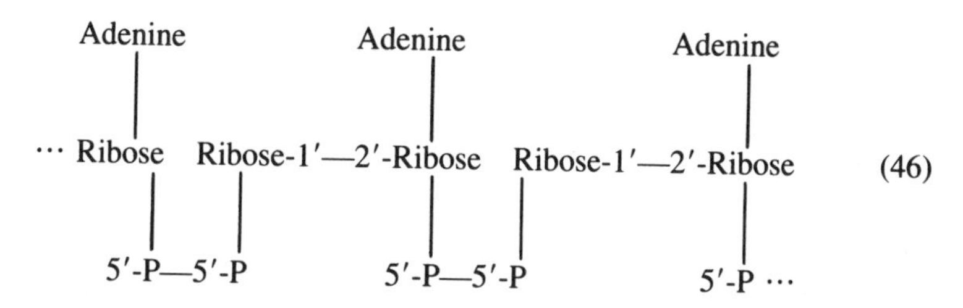

Histidine Biosynthesis

Positions *N*-1 and *C*-2 of ATP are used in the formation of the imidazole ring of histidine. The first steps of this pathway are as follows:

$$\text{ATP} + \text{PP-ribose-P} \rightarrow 1\text{-}(5'\text{-Phosphoribosyl})\text{ATP} \rightarrow \qquad (47)$$

Phosphoribosyl formimino aminoimidazole carboxamide ribonucleotide → →

Imidazole glycerolphosphate + Phosphoribosyl aminoimidazole carboxamide

Adenylation of Proteins

Cells regulate the activity of certain enzymes by adenylating them, that is, by transferring the adenylyl moiety of ATP to the phenolic hydroxyl of certain tyrosine residues. The best studied such enzyme is the glutamine synthetase of enterobacteria, which contains up to 12 adenylyl groups.

FACTORS AFFECTING ATP METABOLISM IN TISSUES AND ANIMALS

It has been seen that the synthesis of ATP can be very much influenced by the exogenous supply of adenine, adenosine, and hypoxanthine and its precursors (Chapter 6). One potential source of such purine bases are nucleosides and nucleotides in the diet, and another is circulating purines; the role of each is considered here. In addition, several hormones can affect the synthesis of ATP.

Diet

Dietary components may potentially influence ATP concentrations in three ways: (1) phosphorylation of adenylate and ADP may be affected by caloric intake and dephosphorylation, by the need to phosphorylate various food constituents; (2) the concentrations of substrates for *de novo* synthesis of purines may be affected by diet; and (3) the purine constituents of the diet, whether at the nucleic acid or low-molecular-weight level, might be converted to ATP and elevate its concentration.

It may be difficult if not impossible to distinguish between the first two points from an experimental point of view, inasmuch as both carbohydrates (PP-ribose-P) and amino acids are required for *de novo* synthesis. It is of interest, however, that a diet that severely retarded the growth of rats was found to have no effect on the ATP content of skeletal muscle.[1] In contrast, another study showed that in rats fasted for 18–40 hours, the rate of purine biosynthesis *de novo* was reduced by approximately 65 percent. These rates partially recovered soon after feeding an amino acid mixture.[2]

The older literature on the effect of diet on purine biosynthesis *de novo* has been reviewed.[3] These studies may be summarized briefly as follows: (1) In uricotelic animals there is a direct relation between the dietary intake of nitrogen and the rate of purine biosynthesis leading to uric acid production and excretion. (2) In ureotelic animals there is some evidence that high protein diets stimulate uric acid or allantoin synthesis; however, this point remains controversial. (3) Though high carbohydrate diets lead to increased uric acid excretion in humans, it is not clear whether this is due to stimulation of purine biosynthesis *de novo* or to an effect on the renal excretion of uric acid.

Recent studies have shown that a dietary deficiency of arginine in rats produces a marked decrease in liver ATP concentrations.[4] The chain of events seems to be as follows: arginine deficiency leads to growth retardation, to increased portal and peripheral blood ammonia, and to elevated concentrations of ammonia and glutamine in liver; the latter observations suggest that the urea cycle is unable to completely channel excess nitrogen into urea biosynthesis.[5] Increased concentrations of ammonia, glutamine, or both then promote the increased synthesis of carbamyl phosphate, and this leads to accelerated pyrimidine biosynthesis *de novo*.[6]

Sufficient excess orotic acid is formed to saturate the orotate phosphoribosyltransferase reaction, leading to a 230 percent increase in the concentrations of pyrimidine nucleotides and urinary excretion of still more orotic acid. The utilization of PP-ribose-P by this accelerated pyrimidine nucleotide synthesis apparently reduces the availability of this essential substrate for purine biosynthesis, and as a result, ATP concentrations are reduced to 24 percent of control values.[4]

The purine constituents of the diet, whether at the nucleic or free nucleotide level, might be converted to ATP. There is no question that the purine content of the diet influences serum uric acid concentrations and urinary uric acid excretion in humans; however, there is no evidence to suggest that cellular ATP concentrations are altered by dietary purines. In early studies, ^{15}N-labeled RNA was fed to rats, and it was found that about 1 percent of the nucleic acid purines of the combined viscera was derived from the dietary material.[7] In other studies, in which nonradioactive[8] or ^{15}N-labeled[9] RNA was fed to human subjects, there were indications that a small amount of dietary purine might have been retained in the tissues; this, of course, could not be tested directly. More recently, it was shown that feeding RNA did not alter the rate of purine biosynthesis *de novo* in humans, and it was concluded that tissue concentrations of inhibitory purine bases or nucleotides were not substantially elevated following ingestion of RNA.[10]

Direct evidence on this subject was obtained by Burridge et al.,[11] who fed three radioactive yeast nucleic acids to mice. One preparation was labeled predominantly in the guanine moiety, and six hours after feeding, only 0.095 percent of the injected radioactivity was found in tissue nucleotides and nucleic acid. When a preparation was used in which adenine and guanine were labeled equally, this figure was 0.26 percent. Small intestine, liver, and skeletal muscle contained most of the radioactive purines, and the dietary nucleic acid adenine was utilized somewhat more efficiently than dietary nucleic acid guanine.

Sonoda and Tatibana[12] have also fed specifically labeled nucleic acids to mice, with generally similar results. Nucleic acid adenine was utilized better than dietary nucleic acid guanine, but in either case the rate of utilization for nucleotide and nucleic acid synthesis was quite low; however, these workers did show that the feeding of high amounts of nucleic acids did lead to more extensive nucleotide formation.

Adenosine triphosphate concentrations in tissues were not measured in these studies, but it seems unlikely that they would have been appreciably increased by the small amounts of dietary purines utilized.

Circulating Purines

Purines that can be used for ATP synthesis circulate in the blood in two ways.

First, hypoxanthine is found in serum in very low concentrations.[13] It may arise in small part from dietary purines but most undoubtedly is formed as an end product of adenine nucleotide and RNA turnover. Because xanthine oxidase is localized mainly in liver and kidney (see Chapter 7), hypoxanthine formed in other tissues must be transported to these tissues in order to be further oxidized. Although it is believed that serum hypoxanthine can also be used for nucleotide synthesis, the quantitative importance of this process is not yet clear.

Second, adenine nucleotides (and perhaps inosinate) are transported among tissues in the erythrocytes. In the latter case, the sequence of events is believed to be as follows:

$$\text{Donor tissue: ATP} \rightarrow \rightarrow \rightarrow \text{Adenosine} \tag{48}$$

$$\text{Adenosine leaves the tissue and enters erythrocyte} \tag{49}$$

$$\text{Erythrocytes in donor tissue: Adenosine} \rightarrow \rightarrow \rightarrow \text{ATP} \tag{50}$$

$$\text{Erythrocytes in acceptor tissue: ATP} \rightarrow \rightarrow \rightarrow \text{Adenosine (+ Inosine)} \tag{51}$$

$$\text{Acceptor tissue: Adenosine} \rightarrow \rightarrow \rightarrow \text{ATP} \tag{52}$$

$$\text{Inosine} \rightarrow \text{Hypoxanthine} \rightarrow \text{Inosinate} \rightarrow \text{ATP} \tag{53}$$

Although there is compelling evidence for the existence of this chain of events in a qualitative way, quantitative data are as yet lacking.[14–16] This route is believed to be important for the maintenance of ATP concentrations in cells that cannot synthesize purines *de novo* (e.g., erythrocytes themselves, some bone marrow cells) and those in which this process cannot supply all the ATP and other purines that are required. In addition, it may be a way to supply metabolizable carbohydrate (in the form of nucleoside ribose) for cells such as porcine erythrocytes that cannot use glucose.[17,18]

Hormones

There is both direct and indirect evidence that a number of hormones have effects on purine metabolism, although this matter has not been explored in a systematic manner. Thus insulin-induced hypoglycemia induces ATP catabolism in rat brain, but this would result from any cause of hypoglycemia.[19] Also, it is well established that circulating uric acid levels are greater in men after puberty than in women before menopause, and this does not seem to be due to sex differences in uric acid excretion. It seems possible that androgens might stimulate inosinate synthesis or that estrogens might reduce its rate, but this area has not been studied.

Studies carried out some years ago provided indirect evidence that thyroid-stimulating hormone, adrenalcorticotropin, and chorionic gonadotropin stimulated glucose-6-phosphate dehydrogenase activity in individual target tissues and increased PP-ribose-P synthesis and purine biosynthesis *de novo*.[3]

More recently, injection of epinephrine, glucagon, and tolbutamide into mice was shown to produce marked elevations in PP-ribose-P concentrations in mouse liver;[20,21] as a result, purine biosynthesis *de novo* was accelerated 2–4.5-fold.[21] Insulin had only a small effect, and hydrocortisone and fluorocortisone had none.[20] The availability of PP-ribose-P for purine biosynthesis *de novo* is also increased when rat liver slices are incubated with epinephrine or glucagon.[22] Whether PP-ribose-P synthesis is accelerated through increased substrate availability following glycogenolysis or by decreased end-product inhibition due to lowered ATP concentrations is not clear.

Finally, the tissue hypertrophy that is produced in the rat renal cortex in experimental diabetes,[23] and that produced in the rat heart by isoproterenol treatment[24] is also associated with increased availability of PP-ribose-P for nucleotide synthesis.

FUNCTIONS OF ATP AND ITS UTILIZATION IN TISSUES AND ANIMALS

Several functions of ATP in specialized tissues should be mentioned, if only briefly.

Exercise

Adenosine triphosphate and creatine phosphate together supply the immediate energy needs for muscle contraction, but no change in the total concentration of ATP can be detected during light or moderate muscular exercise. The concentration of ATP does decrease with more strenuous exercise, however, with maximal depletion of 60–80 percent reported. Adenosine diphosphate concentrations are increased, but there may be also a decrease in total adenine nucleotides; much of the remainder is in the form of inosinate, although circulating and urinary hypoxanthine, inosine, and xanthine also increase and are believed to be derived from this loss in muscular ATP.[25–28]

In further studies of different kinds of muscle fiber in rats, it was observed that stimulation of gastrocnemius muscle led to a 50 percent decrease in ATP and total adenine nucleotides, with an equivalent increase in inosinate. In soleus muscle, however, there was less decrease in ATP, and little of this accumulated as inosinate.[29] Carrying this approach further, it was shown that ATP was broken down to 47, 63, 75, and 89 percent of control values in white vastus lateralis, plantaris, red vastus lateralis, and soleus muscles, respectively, during treadmill running. Inosinate accumulation represented 97, 85, 71, and 33 percent of the loss of ATP in the four muscle types, respectively.[30]

ATP and the Nervous System

Adenosine triphosphate and adenosine derived from it have a variety of effects in and on both the peripheral[31–32] and central[33–35] nervous systems. Peripherally, ATP has been proposed as a neurotransmitter in a class of nerves called *purinergic,* although this concept still remains controversial. Both adenosine and ATP can also modulate peripheral pre- and postsynaptic receptors. Centrally, ATP, adenosine, or both have been studied from several points of views as possible neurotransmitters; as regulators of neural excitability; and for their effects on transmitter release, receptor sensitivity, and behavior.

Regulation of Blood Flow

Among the various factors that influence the coronary vasculature are chemical factors that serve as links between the oxygen needs of cardiac parenchymal tissue and the supply of oxygen through the blood. One of the principal mediators of coronary blood flow regulation in response to cardiac metabolic activity is adenosine.[36–38]

The source of this adenosine is ATP, but vasoactive amounts of adenosine can be formed from changes in ATP concentration that are too small to measure accurately. The extent to which adenosine arises from AMP through 5′-nucleotidase, from S-adenosylhomocysteine; or even from cAMP still is not clear, and there is, in fact, evidence that at least two different adenine nucleotide compartments of the heart serve as precursors for the formation of adenosine; the intercellular or intracellular basis of such compartmentation is not known at present. The compartment that produces the regulatory adenosine may differ under normoxic and hypoxic conditions.[39] Exactly how adenosine elicits relaxation of the coronary vascular smooth muscles is still unknown.

There is also evidence that adenosine may play a role in vasoregulation in skeletal muscle as well;[40] however, this is less certain, and it is possible that it is important under some conditions but not others. For example, adenosine concentrations are elevated during ischemic exercise, but not during exercise under normoxic conditions.[41]

Secretable Storage Pools of ATP

Platelets contain two types of granule, one of which, the "dense granules," contain ATP and ADP as well as serotonin and other constituents.[42] These nucleotides are not in ready equilibrium with the active metabolic pool of ATP, but presumably enter or are inserted into the storage granules at some early stage in the production of platelets. Secretion of serotonin is accompanied by the release of ADP and ATP; however, these are very rapidly cleared from the blood. The function of intragranular ATP may be related to the storage of serotonin.

Adenosine triphosphate is also stored in the vesicles or granules responsible for storing acetylcholine and catecholamines.[33–35] Thus the adrenal medullary chromaffin granules, sympathetic nerve vesicles, and cholinergic vesicles all contain ATP in addition to neurotransmitters; this is discharged, at least in part, during release of the catecholamines or acetylcholine. In contrast with platelets,

ATP can be taken up relatively easily into these granules and vesicles, but it is then lost only rather slowly.

CYCLIC PATHWAYS OF ATP UTILIZATION AND FUNCTION

The metabolism of ATP involves a number of cyclic pathways in which ATP is first utilized and then resynthesized without (necessarily) any net loss of ATP. Although many of these cyclic pathways involve transfer of one or more of the phosphate groups of ATP, followed by rephosphorylation to ATP, only those cyclic pathways that involve the purine ring of ATP, either alone or together with the ribose and phosphate moieties, are considered in this chapter.

Methylation Reactions

Two cycles of ATP metabolism involve S-adenosylmethionine. In the first of these, ATP is first converted to S-adenosylmethionine, and the methyl group is then transferred to a variety of acceptors. The S-adenosylhomocysteine residue is converted to adenosine, and this compound is converted back to adenine nucleotides.

$$\text{ATP} + \text{Methionine} \xrightarrow{Mg^{2+}} \text{S-adenosylmethionine} + P_i + PP_i \tag{54}$$

$$\text{S-adenosylmethionine} + \text{acceptor} \rightarrow \text{S-adenosylhomocysteine} + \text{Methylated acceptor} \tag{55}$$

$$\text{S-adenosylhomocysteine} \rightarrow \text{Homocysteine} + \text{Adenosine} \tag{56}$$

$$\text{Adenosine} + \text{ATP} \rightarrow \text{Adenylate} + \text{ADP} \tag{57}$$

The intracellular concentration of S-adenosylmethionine is dependent on the availability of methionine. Thus incubation of Ehrlich ascites tumor cells with 2 m*M* methionine increased S-adenosylmethionine concentrations twofold,[43] and perfusion of rat livers with methionine led to concentrations of S-adenosylmethionine (300 nmol/g) that were three- to fourfold higher than basal values.[44] In nutritional experiments, the liver contents of this metabolite were approximately doubled when diets containing either methionine or cobalamin were administered.[45]

Attempts have been made to estimate the rate of conversion of S-adenosylmethionine to adenosine plus homocysteine. Hoffman, for example, used periodate-oxidized adenosine to inhibit S-adenosylhomocysteine hydrolase and then measured the accumulation of the substrate of this enzyme in mouse liver *in vivo*.[46] A rate of adenosine production of at least 20 nmol/min per gram of liver was determined, and this may be too low because the accumulation of S-adenosylhomocysteine causes some inhibition of methylation processes and hence of S-adenosylhomocysteine formation from S-adenosylmethionine.

In Ehrlich ascites tumor cells incubated with an inhibitor of adenosine phosphorylation, 1.5 percent of intracellular radioactive ATP was converted to S-adenosylmethionine, S-adenosylhomocysteine, and adenosine per hour, with about 40 percent of this remaining in S-adenosylmethionine.[43]

Polyamine Synthesis

The involvement of S-adenosylmethionine in polyamine synthesis may be described briefly as follows:

$$\text{S-adenosylmethionine} \xrightarrow{CO_2} \text{"Decarboxylated SAM"} \tag{58}$$

$$\begin{matrix}\text{"Decarboxylated SAM" +} \\ \text{Putrescine or Spermidine}\end{matrix} \rightarrow \begin{matrix}\text{Methylthioadenosine +} \\ \text{Spermidine or Spermine}\end{matrix} \tag{59}$$

$$\text{Methylthioadenosine} \rightarrow \begin{matrix}\text{Adenine +} \\ \text{Methylthioribose}\end{matrix} \tag{60}$$

$$\begin{matrix}\text{Adenine} \\ \text{+ PP-ribose-P}\end{matrix} \xrightarrow{Mg^{2+}} \text{Adenylate} + PP_i \tag{61}$$

$$\text{Adenylate + ATP} \rightarrow \text{ATP} \tag{62}$$

The conversion of S-adenosylmethionine to adenine in the course of polyamine synthesis is one of only two routes of adenine synthesis in mammalian cells; the other is the direct cleavage of adenosine and deoxyadenosine (see Chapter 7). Evidence for the operation *in vivo* of one or the other or both pathways first originated from studies of patients with a complete deficiency of adenine phosphoribosyltransferase; in the presence of this enzyme, any free adenine formed is rapidly reused, resulting in undetectable adenine concentrations in tissues, serum, and urine.

One adenine phosphoribosyltransferase–deficient patient excreted 0.13 nmol of adenine per 24 hours;[47] this figure does not include the amount converted to 2,8-dioxyadenine, which tends to precipitate and form urinary stones.[48] To determine whether the excreted adenine was derived from *de novo* synthesis as well as diet, Van Acker et al.[49] studied a patient administered a low-purine diet and found that the urinary excretion of adenine and its metabolites amounted to approximately 0.17 nmol per 24 hours; this was about 25 percent of total urinary purines.

Further information regarding the source of free adenine in cells was based on reports that several cultured cell lines are deficient in methylthioadenosine phosphorylase.[50,51] Kamatani and Carson[51] then showed that methylthioadenosine was excreted by such cell lines and accumulated in the medium at a rate of 0.58–0.70 nmol/hr per milligram of protein. In a later study using a cell line also deficient in adenine phosphoribosyltransferase, it was shown that the rate of adenine production (0.27 nmol/hr per milligram of protein) was similar to that of methylthioadenosine.[52] In addition, conditions that increased or decreased the production of methylthioadenosine had similar effects on that of adenine.

Using a different approach, Henderson and Boldt[43] measured the conversion of intracellular radioactive ATP to S-adenosylmethionine, methylthioadenosine, and adenine in Ehrlich ascites tumor cells incubated so that the reutilization of adenine was prevented; 1.23 percent of ATP radioactivity was converted to these metabolites per hour, with about half of this in S-adenosylmethionine.

Purine Nucleotide Cycle

Lowenstein[53] and his collaborators have proposed the operation of a cycle linking ATP and inosinate and have termed this process "the purine nucleotide cycle"; this is a somewhat infelicitous term, as the particular nucleotides involved are not specified, and it is not the only cycle of purine nucleotide metatolism. The following reactions are involved.

Dephosphosphorylation:

$$\text{ATP} \rightarrow \text{ADP} \rightarrow \text{Adenylate} \tag{63}$$

Deamination:

$$\text{Adenylate} \rightarrow \text{Inosinate} + NH_4^+ \tag{64}$$

Reamination:

$$\begin{matrix}\text{Inosinate} \\ + \text{ aspartate} \\ + \text{ GTP}\end{matrix} \rightarrow \begin{matrix}\text{Adenylosuccinate} \\ + \text{ GDP} + P_i\end{matrix} \rightarrow \begin{matrix}\text{Adenylate} \\ + \text{ Fumarate}\end{matrix} \tag{65}$$

Rephosphorylation:

$$\text{Adenylate} \rightarrow \text{ADP} \rightarrow \text{ATP} \tag{66}$$

It has been proposed that this cycle not only plays a role in purine nucleotide metabolism, but also acts as a pathway of deamination of amino acids in some tissues. This process would proceed as follows:

$$\begin{matrix}\text{Various amino acids} \\ + \text{ Oxaloacetate}\end{matrix} \rightarrow \begin{matrix}\text{Various } \alpha\text{-keto acids} \\ + \text{ Aspartate}\end{matrix} \tag{67}$$

$$\begin{matrix}\text{Aspartate} + \text{Inosinate} \\ + \text{ GTP}\end{matrix} \rightarrow \begin{matrix}\text{Adenylosuccinate} \\ + \text{ GDP} + P_i\end{matrix} \tag{68}$$

$$\text{Adenylosuccinate} \rightarrow \begin{matrix}\text{Adenylate} \\ + \text{ Fumarate}\end{matrix} \tag{69}$$

$$\text{Adenylate} \rightarrow \text{Inosinate} + NH_4^+ \tag{70}$$

$$\begin{matrix}\text{Fumarate} \\ + H_2O\end{matrix} \rightarrow \text{Malate} \underset{NAD^+}{\rightarrow} \begin{matrix}\text{Oxaloacetate} \\ + \text{ NADH} + H^+\end{matrix} \tag{71}$$

The operation of the purine nucleotide cycle has been demonstrated in suitably fortified *extracts* of muscle,[54] kidney,[55] brain,[56] and liver.[57] However, some of the conditions used in the demonstration of this cycle in cell extracts, particularly the premise of rather high concentrations of adenylate, do not seem physiologically realistic.

Attempts have been made to demonstrate the operation of the purine nucleotide cycle in intact mammalian cells *in vitro* through the inhibition of adenylosuccinate synthetase by hadacidin. Thus Crabtree and Henderson[58] compared the conversion of radioactive adenine nucleotides to inosine (derived from inosinate, which does not accumulate) in Ehrlich ascites tumor cells treated with hadacidin. The conversion of radioactivity in inosine was increased 25 percent after a 90-minute incubation with hadacidin, but the total amount of radioactivity involved was only about 4 percent that in total adenine nucleotides. Raivio and Seegmiller[59] made similar observations in studies of cultured human fibroblasts.

More recently, however, evidence has been obtained in studies *in vivo*, that are consistent with the operation of the purine nucleotide cycle in animal tissues under certain conditions. Thus exercise greatly stimulated ammonia production in skeletal muscle and also decreased the concentrations of creatine phosphate and ATP. The decrease in ATP was accounted for by small increases in ADP, adenylate, hypoxanthine, and inosine, but mainly by the accumulation of inosinate. Adenylosuccinate was not detected in resting muscle, but appeared during exercise.[60] In later studies, Aragon and Lowenstein[61] showed that the concentrations of fumarate, malate, and oxaloacetate increased considerably during muscular exercise, as is predicted by the purine nucleotide cycle hypothesis. Hadacidin, which partially inhibited the conversion of inosinate to adenylosuccinate under the conditions used, also partially inhibited the rise in citric acid cycle intermediates. In neither of these studies was there evidence for the operation of the cycle in resting muscle.

Electric shock treatment of rat brain also produced a breakdown in ATP, with a transient rise in the concentration of adenylate, followed by a rise in ammonia and in inosinate. The latter was subsequently metabolized by dephosphorylation to inosine and hypoxanthine and by conversion to adenylosuccinate; adenosine was also formed. The relative importance of adenosine formation and deamination for total ammonia production was not estimated, however, and the time course of adenylosuccinate accumulation and disappearance was more clearly associated with that of adenylate rather than with that of inosinate.[62]

Finally, it has been proposed that the purine nucleotide cycle is an important source of urinary ammonia produced during acidosis (in addition to glutamine). The main basis for this hypothesis is the observation that the concentration of inosinate increased fivefold after 48 hours of acid feeding, whereas that of adenylosuccinate decreased; the activities of adenylosuccinate synthetase and adenylosuccinate lyase also increased.[63]

Turnover of RNA

It is well established that a substantial portion of heterogeneous nuclear RNA breaks down very soon after it is synthesized; likewise, messenger RNAs may have relatively short half-lives, although there is considerable variation. In contrast, the various forms of ribosomal and transfer RNA all have about the same rate of turnover. It is generally believed that a major part of the nucleotide and nucleoside end products of RNA catabolism are reutilized, although the exact pathways followed and their quantitative relationships are far from clear.

Various RNAases exist, with different intracellular localizations and substrate specificities; both 2′(or 3′)-nucleoside monophosphates and 5′-nucleoside monophosphates may be end products.[64,65] If 5′-adenylate were to be produced, its reuse would simply require phosphorylation back to ATP. In contrast, 2′(or 3′)-adenylate would have to be dephosphorylated to adenosine, and the adenosine would then have to be rephosphorylated; in both cases, deamination could also occur. The relative amounts of each type of nucleotide produced and the partition between reuse and further catabolism are not yet firmly established.

Reutilization of Hypoxanthine

The fact that lack of hypoxanthine phosphoribosyltransferase or purine nucleoside phosphorylase activities leads to elevated PP-ribose-P availability and accelerated purine biosynthesis *de novo* indicates that hypoxanthine (or guanine or both) is used for nucleotide synthesis by cells. That is, nucleotide synthesis from this purine base normally appears to compete with the *de novo* pathway for PP-ribose-P; when one pathway is missing, its "share" of PP-ribose-P is available to the other. The hypoxanthine involved might be supplied exogenously, but the major source is believed to be intracellular.

Several attempts have been made to determine the extent to which the hypoxanthine produced in the human body is, in fact, reutilized for nucleotide synthesis rather than being oxidized to xanthine and uric acid. On the basis of measurements of the urinary excretion of purines in normal and hypoxanthine phosphoribosyltransferase deficient children, for example, Murray[66] calculated that approximately 40 mg of hypoxanthine per kilogram of body weight per day was normally reutilized; this amounts to about 85 percent of the total production of this compound.

Using a quite different approach, Edwards et al.[67] came to similar conclusions. They injected radioactive adenine into normal humans and individuals lacking hypoxanthine phosphoribosyltransferase and compared the rate at which radioactivity appeared in urinary uric acid, xanthine, and hypoxanthine. Adenine, of course, cannot be converted to these end products directly, but only after conversions to adenylate and degradation to hypoxanthine. It was found that the excretion of radioactivity rose from 5.6 percent of the amount injected per week in controls to 22.3 percent in patients lacking the ability to convert hypoxanthine to inosinate. From these results, it was calculated that approximately 75 percent of the hypoxanthine normally produced in the body is reutilized rather than immediately oxidized.

ATP Resynthesis in Tissues Following Catabolism

The normal cyclic utilization and resynthesis of ATP described above does not involve changes in ATP concentrations. However, examples can also be given of situations in which ATP catabolism does proceed to the extent that its concentration is lowered, before resynthesis can compensate for this loss. In addition, the effect of such catabolism is sometimes decreased by causing adenine nucleotide synthesis to proceed at accelerated rates. The first situation is found in tissues following anoxia and the second, in the process of extracorporeal blood and tissue preservation.

Response of Synthetic Pathways to ATP Catabolism

One of the first studies of rates of nucleotide synthesis following catabolism was carried out by Marko et al., using kidney slices subjected to anoxia.[68] During postanoxia recovery, it was found that purine biosynthesis *de novo* increased about twofold. In addition, nucleotide synthesis from added radioactive adenine was also increased about 20 percent. This type of recovery was not observed when brain slices were used, however.[69] The same group also first subjected rat hearts to asphyxia or ischemia and then allowed them to recover. Purine biosynthesis *de novo* was then found to increase two- to sixfold, depending on conditions.[70]

Similar observations have been made in liver following loads of fructose (see Chapters 7, 10, and 11).

Blood and Tissue Preservation

For satisfactory transfusion of blood into human patients, erythrocytes should maintain both their structural integrity (i.e., not lyse) and their functional capacity to transport oxygen. One of the important limiting factors for posttransfusion erythrocyte survival is the concentration of ATP, whereas a key factor in satisfactory oxygen transport is the concentration of 2,3-diphosphoglycerate. Over the course of a number of years (for reviews, see Bartlett[71] and Peck et al.[72]), it has been found that the lifetime of ATP during storage can be roughly doubled if adenine, inosine, P_i, and pyruvate (especially the first three) are added to the basic storage solution.

The addition of adenine by itself is not sufficient, for in order to lead to a net synthesis of ATP, it requires (1) PP-ribose-P, which, in turn, needs ribose-5-P, ATP, and P_i and (2) ATP and P_i for the phosphorylation of adenylate to ADP and finally to ATP. Thus there is a presumption that although ATP and even total adenine nucleotide concentrations may be reduced, there is enough adenylate and ADP present to serve as substrates for the ATP-generating reactions of glycolysis.

Stored cells apparently do not use glucose well, either for glycolysis or for pentose phosphate synthesis, and the main function of the inosine that is added is to furnish metabolizable sugar in the form of ribose-5-P. Inosine is thus phosphorolyzed to ribose-1-P in a reaction requiring P_i. This is then converted to ribose-5-P, and a large part of this sugar phosphate is converted to glycolytic intermediates. These, plus the added P_i, stimulate the glycolytic pathway to phosphorylate ADP (and also produce 2,3-diphosphoglycerate). Pyruvate, if present, helps through the lactate dehydrogenase reaction to maintain a supply of NADH for the 3-phosphoglycerate dehydrogenase reaction.

Some of the ribose-5-P that is formed, however, is converted to PP-ribose-P, and this process is stimulated by the high concentration of P_i that is added. Adenine then reacts with this PP-ribose-P to form adenylate, and this is successively phosphorylated to ADP and then ATP.

Adenosine triphosphate is also an important limiting factor for the preservation of kidneys for transplantation, although the situation here is more complex than with blood preservation. Although kidneys can be preserved satisfactorily if chilled, oxygenated, or both, they often are first exposed to a certain period of warm ischemia, during which ATP catabolism can occur. Recent studies have shown, however, that perfusion of kidneys during preservation with solutions of ATP can maintain tissue ATP concentrations at normal levels.[73] It seems likely, however, that it is not ATP itself, but rather some metabolite, that actually enters the cells. Limited catabolism can be compensated for by adenine nucleotide synthesis following restoration of blood circulation; however, extensive catabolism leads to irreversible tissue damage and inability to synthesize purines to the extent that ATP concentrations recover.[74]

REFERENCES

1. R. E. Howarth and R. L. Baldwin, *J. Nutr.* **101,** 485 (1971).
2. P. F. Semple, A. R. Henderson, and J. A. Boyle, *Clin. Sci. Mol. Med.* **46,** 37 (1974).
3. J. F. Henderson, *Regulation of Purine Biosynthesis,* American Chemical Society, Washington, DC (1972).
4. A. S. Hasson and J. A. Milner, *Metabolism* **30,** 739 (1981).
5. J. A. Milner and W. J. Visek, *J. Nutr.* **108,** 382 (1978).
6. A. S. Hasson and J. A. Milner, *Arch. Biochem. Biophys.* **194,** 24 (1979).
7. P. M. Roll, G. B. Brown, J. F. DiCarlo, and A. S. Schultz, *J. Biol. Chem.* **180,** 329 (1949).
8. J. H. Ayvazian and S. Skupp, *J. Clin. Invest.* **45,** 1859 (1966).
9. D. Wilson, A. Beyer, C. Bishop, and J. H. Talbott, *J. Biol. Chem.* **209,** 227 (1954).
10. J. Bowering, D. H. Calloway, S. Margen, and N. A. Kaufman, *J. Nutr.* **100,** 249 (1970).
11. P. W. Burridge, R. A. Woods, and J. F. Henderson, *Can. J. Biochem.* **54,** 500 (1976).
12. T. Sonoda and M. Tatibana, *Biochem. Biophys. Acta* **521,** 55 (1978).
13. G. A. Taylor, P. J. Dady, and K. R. Harrap, *J. Chromatogr.* **183,** 421 (1980).
14. J. F. Henderson and G. A. LePage, *J. Biol. Chem.* **234,** 3119 (1959).
15. M. H. Lerner and B. A. Lowy, *J. Biol. Chem.* **249,** 959 (1974).
16. J. B. Pritchard, N. O'Connor, J. M. Oliver, and R. D. Berlin, *Am. J. Physiol.* **229,** 967 (1975).
17. R. P. Watts, K. Brendel, M. G. Luthra, and H. D. Kim, *Life Sci.* **25,** 1577 (1979).
18. H. D. Kim, R. P. Watts, M. G. Luthra, C. R. Schwabe, R. T. Conner, and K. Brendel, *Biochim. Biophys. Acta* **589,** 256 (1980).

19. B. K. Siesjo, *Brain Energy Metabolism,* Wiley, Chichester, U.K. (1978), p. 380.
20. M. Lalanne and J. F. Henderson, *Can. J. Biochem.* **53,** 394 (1975).
21. S. Brosh, P. Boer, and O. Sperling, *Biomedicine* **35,** 50 (1981).
22. C. Des Rosiers, M. Lalanne, and J. Willemot, *Can. J. Biochem.* **58,** 599 (1980).
23. P. Cortes, C. P. Verghese, K. K. Ventakadhalom, A. M. Schoenberger, and N. W. Levin, *Am. J. Physiol.* **238,** E341 (1980).
24. H.-G. Zimmer and E. Gerlach, *Circ. Res.* **35,** 536 (1974).
25. E. Hultman, J. Bergstrom, and N. M. Anderson, *Scand. J. Clin. Lab. Invest.* **19,** 56 (1967).
26. J. Karlsson, *Acta Physiol. Scand.* Suppl. 358 (1971).
27. S. Rehunen and M. Harkonen, *Scand. J. Clin. Lab. Invest.* **40,** 45 (1980).
28. J. R. Sutton, C. J. Toews, G. R. Ward, and I. H. Fox, *Metabolism* **29,** 254 (1980).
29. R. A. Meyer and R. L. Terjung, *Am. J. Physiol.* **237,** C111 (1979).
30. R. A. Meyer, G. A. Dudley, and R. L. Terjung, *J. Appl. Physiol.* **49,** 1037 (1980).
31. G. Burnstock, *Pharmacol. Rev.* **24,** 509 (1972).
32. G. Burnstock, *J. Exp. Zool.* **194,** 103 (1975).
33. T. W. Stone, *Neuroscience* **6,** 523 (1981).
34. J. W. Phillis and P. H. Wu, *Progr. Neurobiol.* **16,** 187 (1981).
35. N. Morel and F.-M. Meunier, *J Neurochem.* **36,** 1766 (1981).
36. E. Gerlach, J. Schroder, and S. Nees, in *Physiological and Regulatory Functions of Adenosine and Adenine Nucleotides,* H. P. Baer and G. I. Drummond, Eds., Raven, New York (1979), p. 127.
37. J. Schrader, E. Gerlach, and G. Baumann, in *Physiological and Regulatory Functions of Adenosine and Adenine Nucleotides,* H. P. Baer and G. I. Drummond, Eds., Raven, New York (1979), p. 137.
38. R. M. Berne, *Circ. Res.* **47,** 807 (1980).
39. J. Schrader and E. Gerlach, *Pflügers Arch.* **367,** 129 (1976).
40. R. A. Olsson, *Ann. Rev. Physiol.* **43,** 385 (1981).
41. R. D. Phair and H. V. Spacker, *Am. J. Physiol.* **237,** H1 (1979).
42. H. Holmsen and H. J. Weiss, *Annu. Rev. Med.* **30,** 119 (1979).
43. J. F. Henderson and K. L. Boldt, unpublished results (1980).
44. D. R. Hoffman, D. W. Marion, W. E. Cornatzer, and A. Duerre, *J. Biol. Chem.* **255,** 10822 (1980).
45. Y. S. Shin, K. U. Beuhring, and E. L. R. Stokstad, *Mol. Cell. Biochem.* **9,** 97 (1975).
46. J. L. Hoffman, *Arch. Biochem. Biophys.* **205,** 132 (1980).
47. H. A. Simmonds, *Clin. Chim. Acta* **23,** 319 (1969).
48. H. A. Simmonds, K. J. Van Acker, J. S. Cameron, and W. Snedden, *Biochem. J.* **157,** 485 (1976).

49. K. J. Van Acker, H. A. Simmonds, C. Potter, and J. S. Cameron, *New Engl. J. Med.* **297,** 127 (1977).
50. J. I. Toohey, *Biochem. Biophys. Res. Commun.* **78,** 1273 (1977).
51. N. Kamatani and D. A. Carson, *Cancer Res.* **40,** 4178 (1980).
52. N. Kamatani and D. A. Carson, *Biochim. Biophys. Acta 675,* 344 (1981).
53. J. M. Lowenstein, *Physiol. Rev.* **52,** 382 (1972).
54. K. Tornheim and J. M. Lowenstein, *J. Biol. Chem.* **247,** 162 (1972).
55. R. T. Bogusky, L. M. Lowenstein, and J. M. Lowenstein, *J. Clin. Invest.* **58,** 326 (1976).
56. V. Schultz and J. M. Lowenstein, *J. Biol. Chem.* **241,** 485 (1976).
57. K. M. Moss and J. D. McGivan, *Biochemistry* **150,** 275 (1975).
58. G. W. Crabtree and J. F. Henderson, *Cancer Res.* **31,** 985 (1971).
59. K. O. Raivio and J. E. Seegmiller, *Biochim. Biophys. Acta* **299,** 273 (1973).
60. M. N. Goodman and J. M. Lowenstein, *J. Biol. Chem.* **252,** 5054 (1977).
61. J. J. Aragon and J. M. Lowenstein, *Eur. J. Biochem.* **110,** 371 (1980).
62. V. Schultz and J. M. Lowenstein, *J. Biol. Chem.* **253,** 1938 (1978).
63. R. T. Bogusky, K. A. Steele, and L. M. Lowenstein, *Biochem. J.* **196,** 323 (1981).
64. E. A. Barnard, *Annu. Rev. Biochem.* **38,** 677, (1969).
65. M. Sameshina, S. A. Liebhaber, and D. Schlessinger, *Mol. Cell. Biol.* **1,** 75 (1981).
66. A. W. Murray, *Ann. Rev. Biochem.* **40,** 811 (1971).
67. N. L. Edwards, D. Recker, and I. H. Fox, *J. Clin. Invest.* **63,** 922 (1979).
68. P. Marko, E. Gerlach, H.-G. Zimmer, I. Pechan, T. Cremer, and C. Trendelenburg, *Hoppe-Seyler's Z. Physiol. Chem.* **350,** 1669 (1969).
69. E. Gerlach, P. Marko, H.-G. Zimmer, I. Pechan, and C. Trendelenburg, *Experientia* **27,** 876 (1971).
70. H.-G. Zimmer, C. Trendelenburg, H. Kammermeier, and E. Gerlach, *Circ. Res.* **32,** 635 (1973).
71. G. Bartlett, in *The Human Red Cell In Vitro,* T. J. Greenwalt and G. A. Jamieson, Eds., Grune and Stratton, New York (1973), p. 5.
72. C. C. Peck, G. L. Moore, and R. B. Bolin, *CRC Crit. Rev. Clin. Lab. Sci.* **13,** 173 (1981).
73. B. Lytton, V. R. Vaisbort, W. B. Glazier, I. H. Chaudry, and A. E. Baue, *Transplantation* **31,** 187 (1981).
74. C. T. Warnick and H. M. Lazarus, *Can. J. Biochem.* **59,** 116 (1981).

9

Pathological Influences on ATP Metabolism

Adenosine triphosphate concentrations at least potentially may be affected both by inherited abnormalities of one or another of the enzymes of ATP synthesis and catabolism and by nongenetic conditions that alter the concentrations of substrates of these pathways or otherwise accelerate their rates.[1–8] The discussion is divided into two parts, one concerned with pathological influences that at least potentially increase ATP concentrations and the other, with conditions leading to lowered concentrations of ATP. Inherited abnormalities of enzymes of purine metabolism are mentioned at least briefly, even if they do not alter ATP concentrations.

ELEVATED ATP CONCENTRATIONS

Adenosine triphosphate concentrations might logically be expected to increase if the utilization of ATP were retarded or prevented, if its synthesis were stimulated, and if ATP catabolism were decreased, assuming of course, that all other factors remained the same. Although pathological conditions are not known to affect the utilization of ATP, they do both stimulate ATP synthesis and prevent its degradation.

Stimulation of ATP Synthesis

Stimulation of ATP formation might arise as a result of either direct effects on one or another enzyme of ATP synthesis or indirectly by affecting the concentrations of the substrates of these pathways.

No pathological influences are known to directly increase the activities of the enzymes of inosinate biosynthesis *de novo;* of the conversion of inosinate to adenylate; or of adenosine kinase, hypoxanthine phosphoribosyltransferase, or guanylate reductase. However, the activity of adenine phosphoribosyltransferase is elevated up to threefold in erythrocytes of patients with inherited deficiencies of hypoxanthine phosphoribosyltransferase. This is believed to be due to stabilization by the substrate PP-ribose-P, the concentration of which is elevated in this condition.[9]

In contrast, several inherited enzyme deficiencies lead to accelerated inosinate synthesis *de novo* indirectly by altering the synthesis or availability of PP-ribose-P.

PP-Ribose-P Synthetase Hyperactivity

Several human patients with gout have been found to possess abnormal PP-ribose-P synthetases that have greater than normal activity.[10] Such patients have accelerated rates of PP-ribose-P synthesis in erythrocytes and cultured fibroblasts, elevated concentrations of PP-ribose-P *in vitro* and *in vivo,* and considerably increased rates of inosinate synthesis *de novo in vivo*.

Three types of enzyme abnormality have been described. One exhibits decreased sensitivity to end-product inhibition and an increased sensitivity to activation by low concentrations of P_i. A second type has normal kinetic properties, except that the catalytic rate (specific activity per molecule) is increased 2.2-fold. The third type of synthetase has a three- to fourfold higher affinity than normal for ribose-5-P.

Adenosine triphosphate concentrations have not yet been measured in cells with such abnormalities.

Glucose-6-Phosphatase Deficiency

Gout and accelerated purine biosynthesis *de novo* are consequences of the absence of glucose-6-phosphatase, even though this enzyme is apparently only distantly related to purine metabolism; such patients also have glycogen storage disease.[1,4,5,8,10,11] It appears that in the absence of this enzyme, all other pathways of glucose-6-P metabolism are increased, including glycogen synthesis, glycolysis, and the pentose phosphate pathways; therefore, it is presumed that increased availability of PP-ribose-P is the basis for the accelerated inosinate synthesis *de novo* observed in such patients. In addition, parenteral glucagon administration leads to lowered hepatic ATP concentrations and elevations in those of phosphorylated glycolytic intermediates.[12] Thus PP-ribose-P synthesis might also be accelerated as a result of decreased endproduct inhibition. As glucose-6-phosphatase is found only in liver, kidney, and the gastrointestinal tract, it is not surprising that PP-ribose-P concentrations were found to be normal in erythrocytes and cultured fibroblasts from these patients.[13] However, PP-ribose-P concentrations have been reported to be elevated in liver biopsy specimens from such patients.[14]

In this condition, both purine synthesis *de novo* and ATP catabolism are accelerated, and tissue ATP concentrations must thus represent a balance between these influences.

Hypoxanthine Phosphoribosyltransferase Deficiency

Both complete and partial deficiencies of this enzyme are known in humans, and in both cases inosinate synthesis *de novo* is considerably accelerated. Gout is the result of this overproduction of purines in patients with the partial deficiency, whereas those with the complete deficiency not only have problems with uric acid precipitation (mostly kidney stones), but also may have symptoms of cerebral palsy plus choreoathetosis and a unique aggressive behavior toward self as well as others.[2,4,5,7]

Although the basis for the accelerated inosinate biosynthesis observed in patients and cells that lack hypoxanthine phosphoribosyltransferase has been studied for some time, this question is still a matter of controversy. Experimentally, studies have been complicated by the presence of hypoxanthine and of various enzymes of purine catabolism in media commonly used to support the growth of cultured cells.[15,16] The simplest view still is that normally both hypoxanthine phosphoribosyltransferase and amidophosphoribosyltransferase share a limited supply of PP-ribose-P; when the former enzyme is absent, the additional PP-ribose-P is used for inosinate synthesis de novo.[17] Whether one considers this to be increased synthesis of purines or decreased inhibition of *de novo* synthesis by hypoxanthine, the result is the same.

Concentrations of ATP have been found to be normal in cultured human fibroblasts lacking hypoxanthine phosphoribosyltransferase,[17,18] but in enzyme-deficient mouse neuroblastoma cells, they were increased 17 percent.[19] No measurements have been carried out of ATP concentrations *in vivo*.

Decreased ATP Catabolism

A number of inherited deficiencies of enzymes of ATP catabolism have been reported,[6,8] but except in one case, they have not been shown to lead to elevated ATP concentrations.

Adenylate Deaminase Deficiency

Skeletal muscle may lack this enzyme in certain patients with symptoms of easy fatigability and delayed recovery of muscle strength following activity.[20–24] In normal individuals, strenuous exercise leads to accelerated deamination of adenylate, lowered concentrations of ATP, and elevated levels of inosinate; the inosinate then is mostly converted back to adenylate and ATP. Even with less strenuous exercise, this cycle is believed to occur, even though ATP and inosinate concentrations may not change (see Chapter 8). In the absence of adenylate deaminase, adenylate produced as a result of the utilization of ATP for muscular work appears instead to be catabolized by dephosphorylation. The resynthesis of adenylate from the nucleoside and bases thus formed is less efficient than from inosinate, and ATP concentrations might be expected to be reduced until *de novo* synthesis can repair this loss. The clinical relevance of this enzyme abnormality is still unclear, however, as deficient patients may have quite diverse symptoms, and the same symptoms may be found in persons both with and without muscle adenylate deaminase activity.

It has been observed, in fact, that muscle ATP concentrations do decrease following exercise by as much as 94 percent in such patients, and that by 165 minutes thereafter these have recovered to only 67 percent of normal. Following exercise, increased concentrations of adenosine, inosine, and hypoxanthine were noted, but because such catabolites are readily lost from the tissue, the total purine content was also decreased.[25]

Adenosine Deaminase Deficiency

The inherited lack of this enzyme is associated with severe combined immunodeficiency disease in humans.[26–31] This condition is characterized by elevated intracellular concentrations of deoxy-ATP, whereas ATP concentrations in such erythrocytes are reduced; the mechanism involved in not known.[32]

Neither adenosine deaminase patients[33] nor cultured adenosine deaminase-deficient cells[16] synthesized purines *de novo* at accelerated rates.

Purine Nucleoside Phosphorylase Deficiency

The inherited lack of this enzyme also leads to an immunodeficiency disease.[8,29,31] Patients with such a deficiency have been found to synthesize purines *de novo* at accelerated rates, and this is associated with elevated concentrations of PP-ribose-P in their erythrocytes.[34] It is thought that hypoxanthine produced from inosine normally reacts with PP-ribose-P and reduces the availability of this substrate for inosinate synthesis *de novo*. In the absence of hypoxanthine formation, therefore, more PP-ribose-P synthesis is available for *de novo* synthesis. Adenosine triphosphate concentrations have not yet been measured in this condition.

Xanthine Oxidase Deficiency

The inherited lack of this enzyme may result in an asymptomatic state or may lead to the formation of kidney stones containing xanthine.[3] Because this enzyme is not normally present in erythrocytes or leukocytes, few metabolic studies have been carried out. Although increased reutilization of hypoxanthine for nucleotide synthesis might be expected to occur, ATP concentrations have not been measured.

Hereditary High ATP Syndrome

A small number of individuals and families have been reported in which erythrocyte ATP concentrations are elevated to 130–205 percent of normal.[35,36] Clinically, this sometimes is associated with hemolytic anemia. On the basis of other biochemical and clinical abnormalities, it appears that this is not a homogeneous group of individuals, and at least four variant subgroups have been distinguished.

In no case is the biochemical basis of elevated ATP concentration known, but this interesting syndrome certainly deserves further investigation.

LOWERED ATP CONCENTRATIONS

Adenosine triphosphate concentrations might be expected to decrease if the rate of utilization of ATP were reduced, if ATP synthesis were decreased, and if ATP catabolism were increased. Although pathological conditions are not known to affect ATP utilization, they do both decrease ATP synthesis and accelerate ATP catabolism.

Decreased ATP Synthesis

Inherited deficiencies of two enzymes of ATP synthesis are known, but no pathological conditions are known that decrease ATP synthesis by reducing the synthesis or availability of substrates of these pathways.

Adenine Phosphoribosyltransferase Deficiency

Inherited deficiencies of adenine phosphoribosyltransferase have been found in humans.[37–40] Partial (ca. 25 percent of normal activity) deficiency of this enzyme is not uncommon, but complete deficiency is rare; partial deficiency is an asymptomatic state, whereas complete deficiency is characterized by kidney stones composed of 2,8-dioxyadenine. (Adenine formed from S-adenosylmethionine during polyamine synthesis and that cannot be reutilized by the adenine phosphoribosyltransferase reaction can be oxidized by xanthine oxidase to the very insoluble 2,8-dioxyadenine.)

Erythrocyte ATP concentrations are normal both in patients who completely lack this enzyme and in relatives who are heterozygotes.[41] Studies of adenine phosphoribosyltransferase-deficient cultured human fibroblasts showed that PP-ribose-P concentration and rate of ATP synthesis *de novo* were normal.[42]

Hypoxanthine Phosphoribosyltransferase Deficiency

Although the lack of such a synthetic pathway might be considered detrimental, the secondary stimulation of inosinate synthesis *de novo* through increased availability of PP-ribose-P compensates for this deficit.

Increased ATP Catabolism

The greatest number of pathological influences on ATP concentrations are associated with accelerated catabolism of ATP. Although one case of an inherited increase in the activity of a catabolic enzyme is known, many other genetic and nongenetic conditions induce ATP catabolism, mostly by preventing the normal phosphorylation of adenylate and ADP.

Adenosine Deaminase Hyperactivity

A small number of patients have been found who have 45- to 70-fold increases in the activity of adenosine deaminase in erythrocytes.[43] This leads to ATP concentrations that are only about 50 percent of normal, and clinically, to hemolytic anemia. It is not clear, however, whether the lowered ATP concentrations result from increased catabolism of intracellularly produced adenosine (from adenylate or S-adenosylhomocysteine) or from catabolism of adenosine that is supplied to erythrocytes from other tissues, or both.

Trapping of Phosphate

Phosphate trapping leading to increased ATP catabolism may occur in both fructose-1-P aldolase deficiency, glucose-6-phosphatase deficiency, and possibly also in fructose-1,6-bisphosphatase deficiency.[8] In the first abnormal condition mentioned, dietary fructose is phosphorylated but then cannot be metabolized normally through glycolysis. This condition can be mimicked in normal animals simply by injecting massive doses of fructose (see Chapter 8 and 11). In glucose-6-phosphatase and fructose-1,6-bisphosphatase deficiencies, there is an increase in the concentration of phosphorylated sugars, which appears to be the consequence both of decreased dephosphorylation of glucose-6-P and of an increased phosphorylation of free glucose.

Reduced Phosphorylation

The phosphorylation of adenylate and ADP in tissues can be decreased by any condition that deprives cells of oxygen, glucose, another utilizable substrate, or P_i; clearly, a great many such pathological conditions are known.

In brain, ATP concentrations can be decreased by ischemia, hypoxic and anemic hypoxia, severe hypoglycemia, some anaesthetics, and seizures.[44] Adenosine triphosphate catabolism is also produced in cases of acute myocardial infarction and in any other condition that produces cardiac hypoxia.[45,46] There is also evidence for increased ATP breakdown in newborn infants with perinatal complications.[47,48] as well as tissues after acute blood loss[49] or ischemia or hyperthermic stress.[50–52] End products of ATP catabolism also accumulate during human circulatory collapse, smoke inhalation, respiratory acidosis, respiratory distress syndrome, and hypophosphatemia.[8]

Inherited abnormalities of phosphorylation include deficiencies of hexokinase, glucosephosphate isomerase, phosphofructokinase, triosephosphate isomerase, phosphoglycerate kinase, and pyruvate kinase in human erythrocytes.[20] Clinically, these conditions are associated with hemolytic anemia, and in some cases it has been shown that erythrocyte ATP concentrations are decreased.

REFERENCES

1. J. B. Wyngaarden and W. N. Kelley, in *The Metabolic Basis of Inherited Disease*, 3rd ed., J. B. Stanbury, J. B. Wyngaarden, and D. S. Fredrickson, Eds., McGraw-Hill, New York (1972), p. 889.
2. W. N. Kelley and J. B. Wyngaarden, in *The Metabolic Basis of Inherited Disease*, 3rd ed., J. B. Stanbury, J. B. Wyngaarden, and D. S. Fredrickson, Eds., McGraw-Hill, New York (1972), p. 969.
3. J. B. Wyngaarden, in *The Metabolic Basis of Inherited Disease*, 3rd ed., J. B. Stanbury, J. B. Wyngaarden, and D. S. Fredrickson, Eds., McGraw-Hill, New York (1972), p. 992.
4. D. S. Newcombe, *Inherited Biochemical Disorders and Uric Acid Metabolism*, University Park Press, Baltimore, 1975.
5. J. B. Wyngaarden and W. N. Kelley, *Gout and Hyperuricemia*, Grune and Stratton, New York (1976).
6. I. H. Fox, in *Uric Acid*, W. N. Kelley and I. M. Weiner, Eds., Springer, Berlin (1978), p. 93.
7. J. E. Seegmiller, *Ann. Rheum. Dis.* **39,** 103 (1980).

8. I. H. Fox, *Metabolism* **30,** 616 (1981).
9. M. L. Greene, J. A. Boyle, and J. E. Seegmiller, *Science* **167,** 887 (1970).
10. M. A. Becker, in *Uric Acid,* W. N. Kelley and I. M. Weiner, Eds., Springer, Berlin (1978), p. 155.
11. R. R. Howell, in *The Metabolic Basis of Inherited Disease,* 3rd ed., J. B. Stanbury, J. B. Wyngaarden, and D. S. Fredrickson, Eds., McGraw-Hill, New York (1972), p. 149.
12. H. L. Greene, F. A. Wilson, and P. Hefferan, *J. Clin. Invest.* **62,** 321 (1978).
13. M. L. Greene and J. E. Seegmiller, *J. Clin. Invest.* **48,** 32a (1969).
14. R. R. Howell, in *Purine and Pyrimidine Metabolism,* Amsterdam, Elsevier, Amsterdam (1977), p. 353.
15. M. S. Hershfield and J. E. Seegmiller, *J. Biol. Chem.* **252,** 6002 (1977).
16. L. F. Thompson, R. C. Willis, J. W. Stoop, and J. E. Seegmiller, *Proc. Natl. Acad. Sci. (USA)* **75,** 3722 (1978).
17. F. M. Rosenbloom, J. F. Henderson, I. C. Caldwell, W. N. Kelley, and J. E. Seegmiller, *J. Biol. Chem.* **243,** 1166 (1968).
18. D. P. Brenton, K. H. Astrin, M. K. Cruikshank, and J. E. Seegmiller, *Biochem. Med.* **17,** 231 (1977).
19. F. F. Snyder, M. K. Cruikshank, and J. E. Seegmiller, *Biochim. Biophys. Acta* **543,** 556 (1978).
20. W. N. Fishbein, V. W. Arnbrustmacher, and J. L. Griffin, *Science* **200,** 545 (1978).
21. J. B. Shumate, R. Katnik, M. Ruiz, K. Kaiser, C. Frieden, M. H. Brooke, and J. E. Carroll, *Muscle and Nerve* **2,** 213 (1979).
22. R. R. Heffner, *J. Neuropathol. Exp. Neurol.* **39,** 360 (1980).
23. N. C. Kar and C. M. Rearson, *Arch. Neurol.* **38,** 279 (1981).
24. D. J. Hayes, B. A. Summers, and J. A. Morgan-Hughes, *J. Neurol. Sci.* **53,** 125 (1982).
25. R. L. Sabina, J. L. Swain, B. M. Patten, T. Ashizawa, W. E. O'Brien, and E. W. Holmes, *J. Clin. Invest.* **66,** 1419 (1980).
26. S. H. Polmar, *Clin. Haematol.* **6,** 423 (1977).
27. R. Hirschhorn, *Progr. Clin. Immunol.* **3,** 67 (1977).
28. P. Daddona and W. N. Kelley, *Mol. Cell. Biochem.* **29,** 91 (1980).
29. L. F. Thompson and J. E. Seegmiller, *Adv. Enzymol.* **51,** 167 (1980).
30. B. S. Mitchell and W. N. Kelley, *Ann. Intern. Med.* **92,** 826 (1980).
31. K. O. Raivio, *Eur. J. Pediatr.* **135,** 13 (1980).
32. C. M. Smith, A. Belch, and J. F. Henderson, *Biochem. Pharmacol.* **29,** 1209 (1980).
33. G. C. Mills, F. C. Schmalsteig, K. B. Trimmer, A. S. Goldman, and R. M. Goldblum, *Proc. Natl. Acad. Sci. (USA)* **73,** 2867 (1976).

34. I. H. Fox, J. Kaminska, N. L. Edwards, E. Gelfand, K. C. Rich, and W. N. Arnold, *Biochem. Genet.* **18,** 221 (1980).
35. W. N. Valentine and K. R. Tanaka, in *The Metabolic Basis of Inherited Disease,* J. B. Stanbury, J. B. Wyngaarden, and D. S. Fredrickson, Eds., McGraw-Hill, New York (1972), p. 1338.
36. J. P. Bapat and A. J. Baxi, *Biochem. Genet.* **19,** 1017 (1981).
37. W. N. Kelley, R. I. Levy, F. M. Rosenbloom, J. F. Henderson, and J. E. Seegmiller, *J. Clin. Invest.* **47,** 2281 (1968).
38. H. Debray, P. Cartier, A. Temstet, and J. Cendron, *Pediatr. Res.* **10,** 762 (1976).
39. K. J. Van Acker, H. A. Simmonds, C. Potter, and J. S. Cameron, *New Engl. J. Med.* **297,** 127 (1977).
40. W. D. L. Musick, *CRC Crit. Rev. Biochem.* **11,** 1 (1981).
41. B. M. Dean, D. Perrett, H. A. Simmonds, A. Sahota, and K. J. Van Acker, *Clin. Sci. Mol. Med.* **55,** 407 (1978).
42. E. B. Spector, M. S. Hershfield, and J. E. Seegmiller, *Somat. Cell Genet.* **4,** 253 (1978).
43. W. N. Valentine, D. E. Paglia, A. P. Tartaglia, and F. Gilson, *Science* **195,** 783 (1977).
44. B. K. Siesjo, *Brain Energy Metabolism,* Wiley, Chichester, U.K. (1978), passim.
45. T. C. Vary, D. K. Reibel, and J. R. Neely, *Annu. Rev. Physiol.* **43,** 419 (1981).
46. R. B. Jennings, K. A. Reimer, M. L. Hill, and S. E. Mayer, *Circ. Res.* **49,** 892 (1981).
47. H. Manzke, K. Dorner, and J. Grunitz, *Acta Paediatr. Scand.* **66,** 713 (1977).
48. M. C. O'Connor, R. A. Harkness, R. J. Simmonds, and F. E. Hytten, *Br. J. Obst. Gyn.* **88,** 381 (1981).
49. T. V. Kazueva and S. A. Seleznev, *Exp. Biol. Med. Bull.* **89,** 894 (1980).
50. O. D. Saugstad, A. Kroese, H. O. Myhre, and R. Andersen, *Scand. J. Clin. Lab. Invest.* **37,** 517 (1977).
51. O. Osswald, H.-J. Schmitz, and R. Kemper, *Pflügers Arch.* **374,** 45 (1977).
52. P. E. Tuchschmid, U. Boutellier, E. A. Koller, and G. V. Duc, *Pediatr. Res.* **15,** 28 (1981).

10

Effects of Drugs

Adenosine triphosphate concentrations may be both elevated and lowered by a wide variety of drugs, and these effects may be accomplished through a number of different mechanisms. As many of these drugs are used as biochemical "tools" to manipulate ATP concentrations, this subject is of considerable experimental significance. This chapter categorizes the different mechanisms by which drugs either increase or decrease ATP concentrations, and wherever possible, give examples of drugs that act in each category; space does not permit a complete listing and discussion of every drug that affects ATP concentrations, however.

ELEVATION OF ATP CONCENTRATIONS

Adenosine triphosphate concentrations may be at least potentially increased if (1) the utilization of ATP is inhibited, (2) ATP synthesis is stimulated, and (3) ATP catabolism is inhibited.

Inhibition of ATP Utilization

Here, as in Chapter 8, ATP "utilization" refers to the metabolism of the adenosine portion of the ATP molecule, and not to that of the triphosphate portion (Chapter

4). Several types of such utilization may be considered, including RNA synthesis, ribonucleotide reduction leading to deoxyribonucleotides and DNA, the synthesis of various coenzymes derived from ATP, and conversion of adenine nucleotides to guanine nucleotides.

Inhibition of RNA Synthesis

Actinomycin D is a well-known inhibitor of RNA synthesis,[1] but several early studies[2] suggested that it also altered the metabolism of purine nucleotides in a manner that later[3] was shown to be compatible with elevations in ATP and/or GTP concentrations. Direct measurement of nucleotide concentrations in Ehrlich ascites tumor cells treated *in vitro* with relatively low concentrations of actinomycin D showed that ATP concentrations were elevated 30 percent; those of GTP were increased 2.8-fold.[2]

In the same study,[2] a second inhibitor of RNA synthesis was also used, namely, daunomycin; in the tumor cells used, DNA synthesis was also inhibited. This compound had exactly the same (indirect) effects on nucleotide metabolism as did actinomycin D, and although ATP concentrations were not actually determined, it was presumed that they were elevated in this case also.

Several adenosine analogues, including cordycepin, xylosyladenine, and sangivamycin also inhibit RNA synthesis;[4–6] individual analogues have preferential effects on the synthesis of different species of RNA, but this selectivity varies from one tissue to another. This type of effect might also be expected to lead to the accumulation of ATP, but this point has not been explored; furthermore, these compounds also produce an inhibition of PP-ribose-P synthesis and of purine biosynthesis *de novo* that would make any such accumulation difficult to detect.[7]

Inhibition of Ribonucleotide Reduction

Ribonucleotide reductase is believed to be inhibited directly by compounds such as hydroxyurea, guanazole, and aromatic thiosemicarbazones[8,9] and indirectly by conditions that elevate the concentrations of thymidine triphosphate, deoxyadenosine triphosphate, or deoxyguanosine triphosphate.[10] Recent studies have revealed that ATP concentrations are somewhat (ca. 25 percent) elevated in cultured Chinese hamster ovary and human lymphoblast 6410 cells following treatment with growth-inhibitory concentrations of hydroxyurea; however, whether these changes are due solely and directly to inhibition of ribonucleotide reductase remains to be established.[11]

Inhibition of Coenzyme Synthesis

Although ATP is utilized in the synthesis of many coenzymes, the writer is not aware that these processes can be interfered with by pharmacological means. In any case, because the concentrations of these coenzymes are generally small relative to that of ATP, such inhibition would be unlikely to be reflected in any marked change in ATP concentrations.

Inhibition of the Conversion of Adenylate to Guanylate

This process can be inhibited in one or another biological system by (1) a variety of purine derivatives and analogues, (2) imidazole ribonucleoside derivatives and analogues, (3) amino acid analogues, and (4) the unique agent, mycophenolic acid.[12,13] Some of the more recent studies of such drugs provide evidence regarding their secondary effects on ATP synthesis and ATP concentrations.

Inosinate dehydrogenase (EC 1.2.1.14) activity in intact cells has been shown to be inhibited by mycophenolic acid; the inosinate that cannot be used for this reaction in the presence of the drug might be used instead for ATP synthesis, and ATP concentrations might then increase as a consequence. Initial studies using Ehrlich ascites tumor cells *in vitro* showed that mycophenolic acid did not stimulate the conversion of inosinate to adenine nucleotides;[14] apparently, the inosinate that was not oxidized either simply accumulated or was dephosphorylated. In later studies using cultured mouse lymphoma L5178Y cells, ATP concentrations decreased slightly in the first few hours following mycophenolic acid treatment but then returned to control values; they did not become elevated.[15] In contrast, treatment of cultured mouse neuroblastoma cells led to 20 and 30 percent increases in ATP concentrations at different times; however, there was no increase in the conversion of radioactive inosinate to adenine nucleotides.[16] Finally, in mouse leukemia L1210 cells treated *in vivo* with mycophenolic acid, ATP concentrations were unchanged or slightly decreased.[17]

Other inhibitors of inosinate dehydrogenase include several imidazole derivatives and analogues, including virazole,[18] bredinin,[19] aminothiadiazole,[17] and 6-azauridine.[20] Virazole did not stimulate the conversion of inosinate to adenylate in intact tumor cells,[14] but azauridine stimulated this process slightly (12–16 percent) in Chinese hamster ovary cells[20] and more extensively (43–71 percent) in Ehrlich ascites tumor cells.[20] Adenosine triphosphate concentrations were not measured in these studies. Treatment of leukemia L1210 cells with aminothiadiazole did not increase the incorporation of radioactive inosine into ATP and either had no effect on the concentration of ATP or caused it to decrease.[17]

Except in selected cell types, therefore, the inosinate that is not converted to guanine nucleotides in the presence of inhibitors of inosinate dehydrogenase is not redirected toward the synthesis of adenine nucleotides.

Stimulation of ATP Synthesis

Drugs that could stimulate ATP synthesis—whether *de novo,* from adenine or adenosine, or from guanylate or hypoxanthine—all might at least potentially produce elevated concentrations of ATP. Of these possible mechanisms, drugs are known that stimulate purine biosynthesis *de novo* and that block the catabolism of exogenously supplied adenosine and hypoxanthine.

The several drugs that stimulate purine biosynthesis *de novo* appear to do so by increasing the rate of synthesis or availability of PP-ribose-P. This type of drug could, for example, stimulate PP-ribose-P synthesis by accelerating the synthesis of ribose-5-P through the oxidative and nonoxidative pentose phosphate pathways; by increasing the concentration of P_i, which is a required allosteric activator of PP-ribose-P synthetase (EC 2.7.6.1); or by blocking the utilization of PP-ribose-P by other pathways.

Stimulation of the Pentose Phosphate Pathway

The oxidative pentose phosphate pathway can be stimulated by electron acceptors such as methylene blue and phenazine methosulfate,[21] and such treatment has been shown to increase PP-ribose-P concentrations in Ehrlich ascites tumor cells,[22] human skin fibroblasts,[23] and erythrocytes[24,25] and to stimulate nucleotide synthesis from glucose and from purine bases.[26] However, it is not yet known if ATP concentrations actually do increase as a result.

Recently it has been reported that the addition to human erythrocytes of pyrroline-5-carboxylic acid stimulates the oxidative pentose phosphate pathway and accelerated PP-ribose-P synthesis.[27,28] The conversion of this normal metabolite to proline by pyrroline-5-carboxylate reductase generates NADP, which is then used to oxidize glucose-6-P and 6-phosphgluconate.

Decreased Utilization of PP-ribose-P by Alternative Pathways

Interrelationships between purine and pyrimidine nucleotide metabolism have received increasing attention in recent years, and at least some of the interactions that have been observed seem to be due to the fact that both pathways share the common substrate PP-ribose-P.[29]

Several studies have shown that inhibition of pyrimidine biosynthesis *de novo* can result in an acceleration of purine biosynthesis *de novo* and an increase in the concentration of ATP. One such inhibitor is pyrazofurin, whose 5′monophosphate metabolite inhibits orotidylate decarboxylase (EC 4.1.1.23). In cultured mouse lymphoma L5178Y cells treated with pyrazofurin, concentrations of both cytidine triphosphate and uridine triphosphate decrease markedly, whereas those of ATP increase approximately 60 percent; GTP also increases.[30] In similar experiments with cultured Chinese hamster ovary cells, ATP concentrations ranged between 8 and 34 percent above control levels at different times.[31]

A second inhibitor is N-(phosphonoacetyl)-L-aspartate (PALA), which inhibits the second reaction in pyrimidine biosynthesis, aspartate transcarbamylase (EC 2.1.3.2). Treatment of cultured mouse leukemia L1210 and Lewis lung carcinoma cells with PALA markedly reduced pyrimidine ribonucleotide concentrations and at the same time increased the concentration of ATP to approximately 160–180 percent of control values;[32] in another study using Lewis lung carcinoma cells, ATP concentrations increased 225 percent.[33] However, PALA treatment of Chinese hamster ovary cells did not decrease pyrimidine nucleotide concentrations so effectively and had little effect on those of ATP.[31]

Other Drugs Affecting PP-ribose-P

2-Amino-1,3,4-thiadiazole and its 2-ethylamino derivative very effectively stimulate purine biosynthesis *de novo,* although exactly how this is accomplished remains uncertain.[34] In humans, serum uric acid and urinary excretion of uric acid both are elevated, and the incorporation of radioactive glycine into urinary uric acid is increased more than three-fold;[35] the incorporation of radioactive formate and glycine into adenine compounds in rat liver was increased as much as six-fold.[36] The only study in which ATP concentrations actually were measured, however, found either no increase or moderate decreases in ATP, depending on time and dose.[17]

A possible basis for the observed stimulation of purine biosynthesis *de novo* was provided by the observation that ethylaminothiadiazole treatment caused the concentrations of PP-ribose-P in mouse liver to increase as much as twelve-fold.[37] This, in turn, could be due to decreased inhibition of PP-ribose-P synthetase by GTP; aminothiadiazale is an inhibitor of inosinate dehydrogenase,[38] and this, in turn, leads to substantial lowering of the concentrations of this end-product inhibitor.[17]

Decreased Catabolism of Adenosine and Hypoxanthine

These substrates of alternative pathways of ATP synthesis can be metabolized by the catabolic enzymes adenosine deaminase and xanthine oxidase, respectively, and such reactions might lower the effective concentrations of these substrates to the point where not much ATP could be synthesized from them. Because of the widespread presence of adenosine deaminase in animals and cultured cells[39] and of xanthine oxidase in animals, inhibitors of these enzymes are often used when adenosine and hypoxanthine are added exogenously. Because this topic overlaps that of drugs that inhibit ATP catabolism (see below), a fuller consideration of such drugs is deferred until later.

Inhibition of ATP Catabolism

Inhibition of one or another of the reactions of ATP catabolism might lead to elevated ATP concentrations. However, the following conditions would have to be met: (1) the reaction in question should be important for ATP catabolism in the cells being studied; and (2) the cell must be able to convert the compound that accumulates back to ATP.

Initiation of ATP Catabolism

Several drugs have been shown to decrease the rate or extent of ATP catabolism in rat brains subjected to hypoxic conditions; these include indoramin, dihydroergotoxine, and naftidrofuryl.[40] None of these compounds seem to have any effect on ATP under normal conditions, and they appear to act by increasing the phosphorylation of adenylate and ATP and by decreasing ATP utilization. Their mechanism of action is not known more precisely.

Adenylate Deaminase

As already mentioned (Chapter 7), inhibition of adenylate deaminase with high doses of deoxycoformycin almost completely prevented the extensive ATP catabolism in mouse liver *in vivo* that normally follows the administration of fructose;[41] inhibition of this enzyme by deoxycoformycin or coformycin has also been observed by others.[42–44] Unfortunately, really specific inhibitors of this enzyme are not known.

Adenosine Deaminase

Several very potent inhibitors of this enzyme are known, including coformycin, deoxycoformycin, and 9-erythro-(3-hydroxy-2-nonyl)adenine (EHNA).[45,46] These compounds would be expected to retard ATP catabolism only if substantial amounts of adenylate were metabolized by dephosphorylation. It has already been described (Chapter 7) that the catabolism of ATP induced by 2-deoxyglucose and other agents can in some cases be retarded through the use of these compounds.

Dephosphorylation

The conversion of inosinate to inosine is retarded or substantially inhibited under certain conditions used to induce ATP catabolism (e.g., dinitrophenol in Ehrlich ascites tumor cells[47]), but no specific or effective inhibitor of cytoplasmic dephosphorylating enzymes is known. However, *ecto*-5′-nucleotidase can be inhibited by adenosine 5′-(α,β-methylene)diphosphate.[48,49]

Purine Nucleoside Phosphorylase

The inosine analogue, formycin B, can inhibit purine nucleoside phosphorylase under some conditions,[50] but it really is not very potent. Recently, a more active compound, 1-β-D-ribofuranosyl-1,2,4-triazole-3-carboxamide has been described[51] and shown to have a K_i of 5 μM; unfortunately, this compound is metabolized to an inhibitor of inosinate dehydrogenase as well. As already considered, the inosine that might accumulate as a result of inhibition of purine nucleoside phosphorylase cannot directly be reutilized, so these inhibitors would not be expected to affect ATP catabolism.

Xanthine Oxidase

This enzyme is readily inhibited by allopurinol and oxopurinol,[52] and the "reutilization" of at least exogenously supplied hypoxanthine can be considerably increased by pretreatment with allopurinol. However, in some cells, allopurinol can react with PP-ribose-P and thereby lower its availability for purine nucleotide synthesis *de novo*;[53,54] this might cancel any ameliorating effect it might have on ATP catabolism.

Nucleoside Loss From Cells

Nucleosides that are produced in the course of ATP catabolism may leave cells, and their passage across cell membrane is in some cells (and at certain nucleoside concentrations) a mediated process.[55,56] A loss of intracellular nucleosides may, therefore, be at least potentially prevented by inhibitors of this process, such as *p*-nitrobenzylthioinosine or dipyridamol.[55–57]

LOWERING OF ATP CONCENTRATIONS

In general, ATP concentrations may be decreased by drugs that (1) inhibit the pathways of ATP synthesis, (2) increase the rate of utilization of the adenosine moiety of ATP, or (3) increase the rate of ATP catabolism.

Inhibition of ATP Synthesis

The effects of drugs on each route of ATP synthesis are considered separately: the *de novo* pathway and synthesis from adenine, adenosine, hypoxanthine, and guanylate. In addition, the *de novo* pathway is considered in three phases: synthesis of inosinate; conversion of inosinate to adenylate; and the phosphorylation of adenylate. Furthermore, the synthesis of inosinate *de novo* may be inhibited both directly and indirectly; the former type of drug interferes with one or more of the enzymes of this pathway, whereas indirect inhibitors lower the concentration of one or another of its substrates; potent inhibitors are found in each class.

Glutamine Analogues

Several glutamine analogues that contain diazo or other highly reactive groups can irreversibly inhibit the two glutamine-requiring enzymes of the *de novo* synthetic pathway; the most important of these are azaserine and diazo-oxo-norleucine.[58] The latter is the less specific and inhibits both amidophosphoribosyltransferase and phosphoribosylformylglycinamidine synthetase equally well. Azaserine is more selective and preferentially inhibits the former enzyme; however, the latter enzyme can be inhibited at high doses.

The many observations of inhibition of purine biosynthesis *de novo* by azaserine and diazo-oxo-norleucine have been reviewed.[21] In addition, up to 60 percent decreases in the concentration of adenine nucleotides in ascites Sarcoma 180 cells were noted following treatment of tumor bearing mice with azaserine,[59,60] and a 37 percent decrease in rat liver ATP was also found with the use of this drug.[61] In more recent investigations, azaserine was used to produce about 40 percent reductions in the ATP content of Ehrlich ascites tumor cells *in vivo*.[62,63]

Allosteric Inhibition of Amidophosphoribosyltransferase

The pathway of purine biosynthesis *de novo* is readily inhibited by what is usually called "feedback inhibition".[21,64] When this is produced by adding a purine base exogenously, three mechanisms may be responsible: utilization of PP-ribose-P by the added purine base; inhibition of PP-ribose-P synthetase by elevated purine ribonucleotides; or inhibition of the first enzyme of the pathway, amidophosphoribosyltransferase, by the same or related ribonucleotides. When feedback inhibition is exhibited without addition of an exogenous purine base, the latter two mechanisms may be responsible.

Most drugs that inhibit purine biosynthesis appear to do so either by reacting with PP-ribose-P or by inhibiting PP-ribose-P synthetase; these are discussed in the appropriate section below. 6-Methylmercaptopurine ribonucleoside, however, following conversion to its monophosphate derivative, appears specifically to inhibit amidophosphoribosyltransferase, both in cell extracts[65,66] and in at least some intact cells.[67] As demonstration that PP-ribose-P synthetase is not its target, treatment with 6-methylmercaptopurine has been shown either not to alter PP-ribose-P concentrations or to cause them to increase.[67,68] These conclusions may not be true of all cell types, however, as recent reports indicate that in cultured human fibroblasts, 6-methylmercaptopurine ribonucleotide does inhibit PP-ribose-P synthetase.[69,70] In any case, this drug strongly inhibits purine biosynthesis *de novo,* leading to marked decreases in ATP and other purine nucleotides.[71,72]

Glutamine Synthesis

The enzyme glutamine synthetase can be inhibited by methionine sulfoximine, and in one study, this compound was shown to inhibit adenine nucleotide synthesis *de novo* both *in vitro* and *in vivo,* when this process depended on endogenous synthesis of glutamine.[73]

Tetrahydrofolate Coenzyme Synthesis

Adenosine triphosphate biosynthesis *de novo* requires 10-formyl tetrahydrofolate, and the availability of this substrate can be very markedly decreased by treatment with methotrexate and similar compounds.[21,74] These compounds inhibit dihydrofolate reductase (marked with an asterisk) in the following very abbreviated scheme:

5,10-Methylene tetrahydrofolate + Deoxyuridylate	→	Dihydrofolate + Thymidylate	(72)
Dihydrofolate	$\xrightarrow{*}$	Tetrahydrofolate	(73)
Tetrahydrofolate	→	5,10-Methylene tetrahydrofolate	(74)
Tetrahydrofolate	→	10-Formyl tetrahydrofolate	(75)
10-Formyl-tetrahydrofolate	→	Phosphoribosyl formylglycinamide	(76)
10-Formyl-tetrahydrofolate	→	Phosphoribosyl formylaminoimidazole carboxamide	(77)

Although methotrexate and related drugs have been known to inhibit purine biosynthesis *de novo* for many years, it was not until 1975 that ATP concentrations were actually measured following methotrexate treatment.[75] Adenosine triphosphate levels in cultured mouse lymphoma L5178Y cells decreased by approximately 80 percent over the course of eight hours of treatment, whereas those of GTP decreased about 90 percent.

Utilization of PP-Ribose-P

A number of purine base derivatives and analogues utilize PP-ribose-P in the course of being converted to nucleotides; this at least potentially can decrease the availability of this substrate for purine biosynthesis *de novo*.[21]

Decreased P_i Concentrations

Purine and pyrimidine nucleoside analogues that are extensively metabolized by purine nucleoside phosphorylase may indirectly inhibit PP-ribose-P by lowering the intracellular concentration of P_i.[69] This required allosteric activator of PP-ribose-P synthetase is, of course, also a cosubstrate of the phosphorylase.

PP-Ribose-P Synthetase

The enzyme PP-ribose-P synthetase is inhibited by a number of purine and purine nucleoside analogues.[21] One class of such compounds consists of adenosine analogues that are metabolized intracellularly to nucleoside triphosphates; these include cordycepin, xylosyladenine, formycin, nebularin, 7-deazanebularin, and perhaps also tubercidin.[21,76,77] Whether the triphosphates inhibit as analogues of the substrate ATP, or as analogues of ATP when it acts as end-product inhibitor, is not certain.

A second class of nucleoside analogues inhibits PP-ribose-P synthesis without being phosphorylated; this includes psicofuranine, decoyinine, 6-cyclopentylthio-9-hydroxymethylpurine, 9-(3-aminopropyl)adenine, 5′-deoxyadenosine, and perhaps benzylthiopurine.[21,78,79]

Adenylate Synthesis from Inosinate

The enzyme adenylosuccinate synthetase (EC 6.3.4.4) is inhibited by two analogues of its substrate, aspartate. One is hadacidin (N-formyl hydroxyaminoacetic acid[80]), and the other is alanosine (L-2-amino-3-nitroso hydroxylamino propionic acid[81,82]); the latter requires enzymatic activation to its active form, which is the alanosine analogue of phosphoribosyl succinoaminoimidazole carboxamide.[83,84]

When rats were treated with hadacidin alone, liver ATP concentrations were either not affected or lowered only slightly. However, when ATP concentrations were first lowered by use of ethionine, hadacidin then was able to prevent resynthesis from injected inosine.[85] Hadacidin also inhibited adenine nucleoside synthesis from radioactive hypoxanthine in Ehrlich ascites tumor cells *in vitro*[67] and decreased ATP concentrations by about 40–50 percent in these cells *in vivo*.[62,63]

Adenine Phosphoribosyltransferase

No useful inhibitor of this enzyme has yet been identified, but in intact cells, this reaction is, of course, affected indirectly by inhibitors of PP-ribose-P synthesis.

Adenosine Kinase

4-Amino-5-iodo-7-β-D-ribofuranosyl pyrrolo[2,3-d]pyrimidine and 6-N-phenyladenosine have both been used to inhibit adenosine kinase in intact cells.[86–88] Neither is entirely specific, and purine biosynthesis *de novo* is partially inhibited following treatment with both of these compounds. In recent experiments, phenyladenosine was found to have fewer side effects.[89] Inhibitors of adenosine kinase would, of course, not be expected to affect ATP concentrations unless ATP synthesis were made to depend on this pathway by careful manipulation of the other anabolic pathways; this has not yet been reported.

Hypoxanthine Phosphoribosyltransferase

Three compounds have been reported to be potent inhibitors of this enzyme: 6-mercapto-9-(tetrahydro-2-furyl)purine (K_i = 37 μM); 2,6-bis-(hydroxyamino)-9-β-D-ribofuranosylpurine ($K_i = 12$ μM); and 6-iodo-9-tetrahydro-2-furyl)purine ($K_i = 108$ μM).[90] These compounds also partially inhibited the incorporation of radioactive hypoxanthine into nucleotides without affecting PP-ribose-P concentrations. Their possible effects on the synthesis and concentration of ATP under different conditions remain to be tested.

Guanylate Reductase

No inhibitor of this reaction is known.

Increased ATP Utilization

No drugs are known that might lower ATP concentrations by increasing its incorporation into RNA, by increasing the rate of ribonucleotide reduction and ATP incorporation into coenzymes, or by increasing ATP conversion to guanine nucleotides. However, the formation of a "false coenzyme," *S*-adenosylethionine, following administration of ethionine, has been shown to lead to markedly reduced ATP concentrations.[91]

Increased ATP Catabolism

As discussed in Chapter 7, ATP catabolism can be induced by drugs that inhibit one or another process of adenylate and ADP phosphorylation or that serve as phosphate "traps."

ADP Phosphorylation

Adenosine triphosphate catabolism is induced both by inhibitors of glycolysis and of oxidative phosphorylation, if conditions are such as to require the cells to depend on one or the other of these processes. Thus, to refer only to selected recent studies, inhibitors of glycolysis such as iodoacetate, iodoacetamide, and sodium fluoride all stimulate ATP breakdown.[92,93] Similarly, treatment with dinitrophenol, rotenone, antimycin A, oligomycin, and carboxyl cyanide *m*-chlorophenylhydrazine stimulated ATP catabolism in various cell types.[86,87,94]

Phosphate Traps

This category of drugs (including some natural metabolites that in this case may be used pharmacologically) has already been referred to in several contexts and thus does not require extensive discussion here. Such compounds act by forcing rapid or extensive phosphorylation and hence utilization of the γ phosphate of ATP; in some cases the phosphate is metabolically irrecoverable, whereas in others it can eventually be used again for phosphorylation of ADP. These traps may be divided into several classes: unmetabolizable sugars such as deoxyglucose[86,87,92,93,95] and deoxygalactose;[96,97] metabolizable sugars such as fructose[98–100] and glucose;[101,102] and some purine ribonucleoside derivatives and analogues.[92]

Drugs of Unknown Mechanism of Action

In recent years, a number of other drugs have also been identified as inducers of ATP catabolism, although their precise sites of action are not known. These include ethidium,[103] isometamidium,[103] bikaverin,[104] dactylarin,[105] misonidazole,[106] isoproterenol and hydralazine,[107] and other related substances. Each produces a distinctive pattern of accumulation of intermediates of ATP catabolism and may prove to be a useful "tool" for dissecting further details of this process.

REFERENCES

1. I. H. Goldberg, in *Antieoplastic and Immunosuppressive Agents Part II,* A. C. Sartorelli and D. G. Johns, Eds. Springer, Berlin (1975), p. 582.
2. F. F. Snyder and J. F. Henderson, *Can. J. Biochem.* **52,** 263 (1974).
3. F. F. Snyder and J. F. Henderson, *Can. J. Biochem.* **51,** 943 (1973).
4. R. I. Glazer, *Toxicol. Appl. Pharmacol.* **46,** 191 (1978).
5. R. I. Glazer, T. J. Lott, and A. L. Peale, *Cancer Res.* **38,** 2233 (1978).
6. R. I. Glazer, K. D. Hartman, and O. J. Cohen, *Biochem. Pharmacol.* **30,** 2697 (1981).
7. J. F. Henderson and G. Zombor, unpublished results (1982).
8. I. H. Krakoff, in *Antineoplastic and Immunosuppressive Agents Part II,* A. C. Sartorelli and D. G. Johns, Eds. Springer, Berlin (1975), p. 789.
9. K. C. Agrawal and A. C. Sartorelli, in *Antineoplastic and Immunosuppressive Agents Part II,* A. C. Sartorelli and D. G. Johns, Eds. Springer, Berlin (1975), p. 793.
10. L. Thelander and P. Reichard, *Ann. Rev. Biochem.* **48,** 133 (1979).
11. J. Hordern and J. F. Henderson, unpublished results (1982).
12. J. F. Henderson, in *Uric Acid* (*Handbook of Experimental Pharmacology,* vol. 51), W. N. Kelley and I. M. Weiner, Eds., Springer, Berlin, (1978), p. 75.
13. C. A. Nichol, in *Antineoplastic and Immunosuppressive Agents Part II,* A. C. Sartorelli and D. G. Johns, Eds., Springer, Berlin (1975), p. 434.
14. C. M. Smith, L. J. Fontenelle, H. Muzik, A. R. P. Paterson, H. Unger, L. Brox, and J. F. Henderson, *Biochem. Pharmacol.* **23,** 2727 (1974).
15. J. K. Lowe, L. Brox, and J. F. Henderson, *Cancer Res.* **37,** 736 (1977).

16. C. E. Cass, J. K. Lowe, J. M. Manchak, and J. F. Henderson, *Cancer Res.* **37,** 3314 (1977).
17. J. A. Nelson, L. M. Rose, and L. L. Bennett, Jr., *Cancer Res.* **36,** 1375 (1978).
18. D. G. Streeter, J. T. Witkawski, G. P. Knare, R. W. Sidwell, R. J. Bauer, R. K. Robins, and L. N. Simon, *Proc. Natl. Acad. Sci. (USA) 70, 1174 (1973).*
19. K. Sakaguchi, M. Tsujino, M. Yoshizawa, K. Mizuno, and K. Hayano, *Cancer Res.* **35,** 1643 (1975).
20. D. Hunting, G. Zombor, and J. F. Henderson, *Biochem. Pharamcol.* **29,** 2261 (1980).
21. J. F. Henderson, *Regulation of Purine Biosynthesis,* American Chemical Society, Washington, DC (1972).
22. J. F. Henderson and M. K. Y. Khoo, *J. Biol. Chem.* **240,** 2349 (1965).
23. W. N. Kelley, I. H. Fox, and J. B. Wyngaarden, *Clin. Res.* **18,** 457 (1970).
24. A. Hershko, A. Razin, and J. Mager, *Biochim. Biophys. Acta* **184,** 64 (1969).
25. A. Hershko, A. Razin, T. Shoshani, and J. Mager, *Biochim. Biophys. Acta* **149,** 59 (1967).
26. J. F. Henderson and M. K. Y. Khoo, *J. Biol. Chem.* **240,** 2358 (1965).
27. G. C. Yeh and J. M. Phang, *Biochem. Biophys. Res. Commun.* **94,** 450 (1980).
28. G. C. Yeh and J. M. Phang, *Biochem. Biophys. Res. Commun.* **103,** 118 (1981).
29. M. Tatibana, in *Uric Acid,* W. N. Kelley and I. M. Weiner, Eds., Springer, Berlin (1978), p. 125.
30. E. C. Cadman, D. E. Dix, and R. E. Handschumacher, *Cancer Res.* **38,** 682 (1978).
31. D. Hunting and J. F. Henderson, *Biochem. Pharmacol.* (1982), 31, 1109.
32. J. D. Moyer and R. E. Handschumacher, *Cancer Res.* **39,** 3089 (1979).
33. T. W. Kensler, G. Mutter, J. G. Hankerson, L. J. Reck, C. Harley, N. Han, B. Ardalan, R. L. Cysyk, R. K. Johnson, H. N. Jayaram, and D. A. Cooney, *Cancer Res.* **41,** 894 (1981).
34. D. L. Hill, *Cancer Chemother. Pharmacol.* **4,** 215 (1980).
35. I. H. Krakoff and M. E. Balis, *J. Clin. Invest.* **38,** 907 (1959).
36. L. Shuster and A. Goldin, *Biochem. Pharmacol.* **2,** 17 (1959).
37. M. Lalanne and J. F. Henderson, *Can. J. Biochem.* **53,** 394 (1975).
38. J. A. Nelson, L. M. Rose, and L. L. Bennett, Jr., *Cancer Res.* **37,** 182 (1977).
39. J. F. Henderson, in *Physiological and Regulatory Functions of Adenosine and Adenine Nucleotides,* H. P. Baer and G. I. Drummond, Eds., Raven, New York (1979), p. 315.
40. M. G. Wyllie, P. M. Paciorek, and J. F. Waterfall, *Biochim. Pharmacol.* **30,** 1605 (1981).
41. D. Siepen and J. F. Henderson, unpublished results (1981).
42. R. P. Agarwal and R. E. Parks, Jr., *Biochem. Pharmacol.* **26,** 663 (1977).

43. G. Van den Berghe, F. Bontemps, and H.-G. Hers, *Biochem. J.*, **188,** 913 (1980).
44. M. Debatisse, M. Berry, and G. Buttin, *J. Cell. Physiol.* **106,** 1 (1981).
45. R. I. Glazer, *Cancer Chemother. Pharmacol.* **4,** 227 (1980).
46. J. F. Henderson, L. Brox, G. Zombor, D. Hunting, and C. A. Lomax, *Biochem. Pharmacol.* **26,** 1967 (1977).
47. C. A. Lomax, A. S. Bagnara, and J. F. Henderson, *Can. J. Biochem.* **53,** 231 (1975).
48. R. A. Olsson, M. K. Gentry, and R. S. Townsend, *Current Topics in Coronary Research,* vol. 39, Plenum, New York (1973), p. 27.
49. M. K. Gentry and R. A. Olsson, *Anal. Biochem.* **64,** 624 (1975).
50. M. B. Sheen, B. K. Kim, and R. E. Parks, Jr., *Mol. Pharmacol.* **4,** 293 (1968).
51. R. C. Willis, R. K. Robins, and J. E. Seegmiller, *Mol. Pharmacol.* **18,** 287 (1980).
52. G. B. Elion, in *Uric Acid,* W. N. Kelley and I. M. Weiner, Eds., Springer, Berlin (1978), p. 485.
53. I. H. Fox, J. B. Wyngaarden, and W. W. Kelley, *New Engl. J. Med.* **283,** 1177 (1970).
54. W. N. Kelly and J. B. Wyngaardern, *J. Clin. Invest.* **49,** 602 (1970).
55. A. R. P. Paterson, in *Physiological and Regulatory Functions of Adenosine and Adenine Nucleotides,* H. P. Baer and G. I. Drummond, Eds., Raven, New York (1979), p. 305.
56. A. R. P. Paterson, *Pharmacol. Therap.* **12,** 515 (1981).
57. W. Schutz, J. Schrader, and E. Gerlach, *Am. J. Physiol.* **240,** H963 (1981).
58. L. L. Bennett, Jr., in *Antineoplastic and Immunosuppressive Agents Part II,* A. C. Sartorelli & D. G. Johns, Eds. Springer, Berlin (1975), p. 484.
59. A. C. Sartorelli and B. A. Booth, *Mol. Pharmacol.* **3,** 71 (1967).
60. G. T. Cramer and A. C. Sartorelli, *Biochem. Pharmacol.* **18,** 1355 (1969).
61. D. E. Hyams, E. B. Taft, G. D. Drummey, and K. J. Isselbacher, *Lab. Invest.* **16,** 604 (1967).
62. C. M. Smith and J. F. Henderson, *Can. J. Biochem.* **54,** 341 (1976).
63. J. Barankiewicz and J. F. Henderson, *Can. J. Biochem.* **55,** 257 (1977).
64. E. W. Holmes, in *Uric Acid,* W. N. Kelley and I. M. Weiner, Eds. Springer, Berlin (1978), p. 21.
65. D. L. Hill and L. L. Bennett, Jr., *Biochemistry* **8,** 122 (1969).
66. B. S. Tay, R. M. Lilley, A. W. Murray, and M. R. Atkinson, *Biochem. Pharmacol.* **18,** 936 (1969).
67. J. F. Henderson and M. K. Y. Khoo, *J. Biol. Chem.* **240,** 3104 (1965).
68. A. R. P. Paterson and M. C. Wang, *Cancer Res.* **30,** 2379 (1970).
69. G. Planet and I. H. Fox, *J. Biol. Chem.* **251,** 5839 (1976).

70. R. C. K. Yen, K. O. Raivio, and M. A. Becker, *J. Biol. Chem.* **256,** 1839 (1981).
71. C. T. Warnick and A. R. P. Paterson, *Cancer Res.* **33,** 1711 (1973).
72. R. A. Woods, R. M. Henderson, and J. F. Henderson, *Eur. J. Cancer* **14,** 765 (1978).
73. L. J. Fontenelle and J. F. Henderson, *Biochim. Biophys. Acta* **177,** 88 (1969).
74. J. R. Bertino, in *Antineoplastic and Immunosuppressive Agents Part II,* A. C. Sartorelli and D. G. Johns, Eds., Springer, Berlin (1975), p. 468.
75. W. M. Hryniuk, L. W. Brox, J. F. Henderson, and T. Tamaoki, *Cancer Res.* **35,** 1427 (1975).
76. J. F. Henderson, A. R. P. Paterson, I. C. Caldwell, and M. Hori, *Cancer Res.* **27,** 715 (1967).
77. C. M. Smith, F. F. Snyder, L. J. Fontenelle, and J. F. Henderson, *Biochem. Pharmacol.* **23,** 2023 (1974).
78. C. M. Smith, L. J. Fontenelle, M. Lalanne, and J. F. Henderson, *Cancer Res.* **34,** 463 (1974).
79. D. Hunting and J. F. Henderson, *Biochem. Pharmacol.* **27,** 2163 (1978).
80. H. T. Shigeura and C. N. Gordon, *J. Biol. Chem.* **237,** 1932 (1962).
81. G. R. Gale and G. B. Schmidt, *Biochem. Pharmacol.* **17,** 363 (1968).
82. G. R. Gale, W. E. Ostrander, and L. M. Atkins, *Biochem. Pharmacol.* **17,** 1823 (1968).
83. A. K. Tyagi and D. A. Cooney, *Cancer Res.* **40,** 4390 (1980).
84. A. K. Tyagi, D. A. Cooney, H. N. Jayaram, J. K. Swiniarski, and R. K. Johnson, *Biochem. Pharmacol.* **30,** 915 (1981).
85. K. H. Shull and S. Villa-Trevino, *Biochem. Biophys. Res. Commun.* **16,** 101 (1964).
86. C. A. Lomax and J. F. Henderson, *Cancer Res.* **33,** 2825 (1973).
87. C. A. Lomax, A. S. Bagnara, and J. F. Henderson, *Can. J. Biochem.* **53,** 231 (1975).
88. A. Y. Divekar and M. T. Hakala, *Mol. Pharmacol.* **7,** 663 (1971).
89. J. F. Henderson and K. L. Boldt, unpublished results (1980).
90. A. L. Jadhav, L. B. Townsend, and J. A. Nelson, *Biochem. Pharmacol.* **28,** 1057 (1979).
91. M. T. Vogt and E. Farber, *Arch, Biochem. Biophys.* **141,** 162 (1970).
92. J. F. Henderson, G. Zombor, P. W. Burridge, G. Barankiewicz, and C. M. Smith, *Can. J. Biochem.* **57,** 873 (1979).
93. R. C. Smith, *Comp. Biochem. Physiol.* **69B** 505 (1981).
94. S. S. Matsumoto, K. O. Raivio, and J. E. Seegmiller, *J. Biol. Chem.* **254,** 8956 (1979).
95. K. Letnansky, *Biochem. Z.* **341,** 74 (1964).

96. J. J. Starling and D. O. R. Keppler, *Eur. J. Biochem.* **80,** 373 (1977).
97. H. Lattke, H. K. Koch, R. Lesch, and D. O. R. Keppler, *Virchows Arch. B Cell. Pathol.* **30,** 297 (1979).
98. M. Chen and R. L. Whistler, *Adv. Carbohydr. Chem. Biochem.* **34,** 285 (1977).
99. G. Van den Berghe, *Curr. Top. Cell Regul.* **13,** 97 (1978).
100. I. H. Fox, *Metabolism* **30,** 616 (1981).
101. K. K. Lomberg-Holm, *Biochim. Biophys. Acta* **35,** 464 (1959).
102. E. L. Coe, *Biochim. Biophys. Acta* **118,** 495 (1966).
103. J. F. Henderson, M. L. Battell, G. Zombor, and M. K. Y. Khoo, *Cancer Res.* **37,** 3434 (1977).
104. J. F. Henderson, M. L. Battell, G. Zombor, J. Fuska, and P. Nemec, *Biochem. Pharmacol.* **26,** 1973 (1977).
105. J. F. Henderson, G. Zombor, and M. Miko, *Biochem. Pharmacol.* **29,** 2155 (1980).
106. J. F. Henderson and G. Zombor, *Biochem. Pharmacol.* **29,** 2533 (1980).
107. G. R. N. Jones, R. Ellis, and M. J. N. Frohn, *Oncodevelop. Biol. Med.* **2,** 155 (1981).

INDEX

Acetyl-coenzyme A carboxylase, effect of phosphorylation, 61
Acetyl-coenzyme A synthase, site of ATP bond cleavage, 10
Acetyl phosphate, free energy of hydrolysis, 13-14
Actinomycin D, inhibition of RNA synthesis, 146
Actomyosin, ATP hydrolysis, 46-49
Adenine:
  conversion to adenylate, 72-73, 75
  regulation of ATP synthesis, 84, 90
Adenine phosphoribosyltransferase:
  deficiency disease, 126, 140
  inhibitors, 80
  kinetics, 80
  mammalian cells lacking, 77
  reaction catalyzed, 72
  regulation by adenine, 84
  regulation of ATP synthesis, 80
Adenosine:
  catabolism, 95-96
    deamination, 84-85, 106
    drugs decreasing, 150
  circulating, 120
  conversion to adenylate, 72-73, 75
  formation from ATP, 94-95
  phosphorylation, 84-85, 106
  regulation of ATP catabolism, 106
  regulation of ATP synthesis, 84-85, 90
  regulation of coronary blood flow, 123
  synthesis, 85
Adenosine deaminase, 104
  deficiency disease, 139
  hyperactivity, 141
  inhibitors, 101, 104, 151
  reaction catalyzed, 95
  regulation of ATP catabolism, 106
  tissue distribution, 105
Adenosine diphosphate:
  catabolism in extracellular space, 110-111
  phosphorylation, inhibitors, 157
  regulation of ATP synthesis, 75
  transport system of mitochondria, 59-60
Adenosine kinase:
  inhibitors, 81, 156
  kinetics, 80
  mammalian cells lacking, 77
  reaction catalyzed, 72
  regulation of ATP catabolism, 106
  regulation of ATP synthesis, 80-81
  tissue distribution, 105
Adenosine phosphorylase, reaction catalyzed, 96
Adenosine phosphosulfate, synthesis, 117
Adenosine triphosphatase, site of ATP bond cleavage, 10
Adenosine triphosphatase (Na^+ -K^+):
  functions, 41-42
  inhibitors, 42
  localization, 42
  mechanism of ATP hydrolysis, 42-45
  membrane orientation, 45-46
  phosphoenzyme, 42-45
    membrane orientation, 45-46
  substrate specificity, 42
  subunit structure, 45
Adenosine triphosphate:
  in adenylation of proteins, 118
  ATP:deoxyATP ratios, 115
  catabolism:
    cleavage of glycosidic bond, 95-96
    degradation of purine ring, 96-97
    dephosphorylation, 94-95
    in extracellular space, 110-111
    functions of, 97-98
    initiation of, 98-100
    in intact cells, 108-110
    rate-limiting step, 101-103
    regulation of alternative pathways, 100-103
    relative rates of alternative pathways, 101-103, 108-110
    resynthesis following, 130-132

conservation, 98
discovery, 1-4
hydrolysis, standard free energy, 10-15
incorporation into RNA, 115
intracellular compartmentalization, 5
intracellular concentration, 4-5
elevated, 135-140, 145-152
lowered, 140-142, 152-158
maintenance of, 58-60
metabolism:
effect of circulating purines, 120-121
effect of diet, 118-120
effect of hormones, 121
in methylation reactions, 124-125
structure, 2
synthesis, 3-4
from adenylate, 72-73, 75
coordination of alternative pathways, 75-79
de novo, 69-75, 118
rate-limiting steps, 76-87
relationships between alternative pathways, 88-90
response to substrate concentration, 88
in synthesis of coenzymes, 116
in synthesis of guanylate nucleotides, 116
in synthesis of histidine, 117
in synthesis of poly ADP ribose, 117
transport system of mitochondria, 59-60
turnover time, 98
utilization, 3-4
Adenosine triphosphate-citrate lyase, effect of phosphorylation, 61
Adenosine-5′-(α,β-methylene)diphosphate, inhibition of *ecto*-5′-nucleotidase, 151
Adenylate:
catabolism, 94-95
deamination, 86
dephosphorylation, 86
in extracellular space, 110-111
conversion to guanylate, inhibitors, 147-148
formation in ATP catabolism, 94-95
metabolism, 106
phosphorylation, 72, 86
regulation of ATP catabolism, 106
regulation of ATP synthesis, 75, 86
synthesis:
from adenine, 72-73, 75
from adenosine, 72-73, 75
from inosinate, 71, 86-87
Adenylate deaminase:
activation, 58-59
changes in total activity, 104
deficiency disease, 138-139
inhibitors, 101, 104, 109, 151
in intact cells, 108-109
reaction catalyzed, 95
regulation of ATP catabolism, 104
role in maintenance of intracellular ATP concentration, 58-59
stimulation, 104
Adenylate kinase:
reaction catalyzed, 94
role in maintenance of intracellular ATP concentration, 58-59
Adenylate nucleotidase, tissue distribution, 105
Adenylosuccinate lyase:
inhibitors, 80
kinetics, 80
reaction catalyzed, 71
regulation of ATP synthesis, 75, 80
Adenylosuccinate synthetase:
inhibitors, 80, 128, 155-156
kinetics, 80
reaction catalyzed, 71
regulation by aspartate, 83
regulation of ATP synthesis, 75, 80
Adenylyl transferases, site of ATP bond cleavage, 10
ADP, *see* Adenosine diphosphate
Adrenalcorticotropin, effect on ATP metabolism, 121
Alanosine, inhibition of adenylosuccinate synthetase, 156-157
Allopurinol, inhibition of xanthine oxidase, 152
Amidophosphoribosyltransferase:
inhibitors, 79
allosteric, 153-154
end product, 89
kinetics, 79
molecular weight, 79
reaction catalyzed, 70
regulation of ATP synthesis, 79, 88
Amino acids, deamination, in purine nucleotide cycle, 127
Amino acyl-tRNA synthases, site of ATP bond cleavage, 10
9-(3-Aminopropyl)adenine, inhibition of PP-ribose-P synthetase, 155

Aminothiadiazole, inhibition of inosinate dehydrogenase, 147
2-Amino-1,3,4-thiadiazole, effect on PP-ribose-P concentration, 149
Androgens, effect on ATP metabolism, 121
Animal cells:
  ATP:adenylate ratio, 5
  ATP:ADP ratio, 5
  ATP concentration, 5
  ATP:deoxy-ATP ratio, 5
  ATP synthesis, 77
Anoxia, recovery from, purine biosynthesis, 130-131
Antimycin A, effect on ATP catabolism, 157
Arginine, dietary, effect on ATP metabolism, 119
Ascites tumor cells, ATP catabolism, 102
Asparagine, regulation of ATP synthesis, 83
Aspartate, regulation of ATP synthesis, 75, 83
Aspartate transcarbamylase, inhibitors, 149
ATP, *see* Adenosine triphosphate
ATPase, *see* Adenosine triphosphatase
Azaserine, inhibition of ATP synthesis, 153
6-Azauridine, inhibition of inosinate dehydrogenase, 147

Bacteria, BF_1-ATPase, 30
BF_1-adenosine triphosphatase, 30
Bikaverin, effect on ATP catabolism, 158
2,6-Bis-(hydroxyamino)-9-β-D-ribofuranosylpurine, inhibition of hypoxanthine phosphoribosyltransferase, 156
Blood:
  coronary flow, effect of adenosine, 123
  preservation, 131-132
Bone marrow cells, ATP concentration, 121
Brain:
  ATP catabolism, 102
  ATP concentration, 142

Calcium, complex with ATP:
  coordination site, 15-18
  dissociation constants, 15-18
Carbohydrates, dietary, effect on ATP metabolism, 118-119
Carboxyl cyanide *m*-chlorophenylhydrazine, effect on ATP catabolism, 157
Cerebral palsy, 137
CF_1-adenosine triphosphatase, 30
Chemiosmotic hypothesis, for electron transport-linked ATP synthesis, 30-31
Chloroplasts, CF_1-ATPase, 30
Cholinergic vesicles, ATP storage, 123
Chorionic gonadotropin, effect on ATP metabolism, 121
Chromaffin granules, ATP storage, 123
Coenzyme A, synthesis, 117
Coformycin, inhibition of adenosine deaminase, 101, 151
Contractile systems, ATP-dependent, 46-49
Cordycepin:
  inhibition of PP-ribose-P synthetase, 155
  inhibition of RNA synthesis, 146
Cyclic adenosine monophosphate, protein kinase cascade system, 60
3′,5′-Cyclic adenylate, synthesis, 116
6-Cyclopentyl-thio-9-hydroxymethylpurine, inhibition of PP-ribose-P synthetase, 155
Cytosol, ATP concentration, 5
  maintenance of, 59-60

Dactylarin, effect on ATP catabolism, 158
Daunomycin, inhibition of RNA synthesis, 146
Decoyinine, inhibition of PP-ribose-P synthetase, 155
Deoxyadenosine:
  degradation, 96
  inhibition of PP-ribose-P synthetase, 155
Deoxyadenosine diphosphate, metabolism, 115
Deoxyadenosine triphosphate:
  ATP:deoxyATP ratios, 115
  formation, 115
  incorporation into DNA, 115
Deoxycoformycin, inhibition of adenosine deaminase, 101, 151
Deoxygalactose, effect on ATP catabolism, 158
Deoxyglucose, effect on ATP catabolism, 100, 158
Deoxyribonucleic acid, synthesis, 115
Diazo-oxo-norleucine, inhibition of ATP synthesis, 153
Diet, effect on ATP metabolism, 118-120
Dihydroergotoxine, effect on ATP catabolism, 150
Dihydrofolate reductase, inhibitors, 154

Dinitrophenol, effect on ATP catabolism, 99, 157
1,3-Diphosphoglycerate, free energy of hydrolysis, 13-14
Dipyridamol, and nucleoside loss from cells, 152

Ecto enzymes, of ATP catabolism, 110-111
Ecto-5′-nucleotidase, inhibitors, 151
Electron transport:
  and ATP synthesis, 29-37
  chemiosmotic hypothesis, 30-31
Epinephrine, effect on ATP metabolism, 121
Erythro-9-(2-hydroxy-3-nonyl)adenine, inhibition of adenosine deaminase, 101, 151
Erythrocytes:
  ATP catabolism, 103, 105
  ATP concentration, 5
    maintenance of, 121
  ATP synthesis, 105
  ATP turnover, 98
  circulation of purines, 120
  posttransfusion survival, 131
Escherichia coli, succinyl-coenzyme A synthetase, 28
Ethidium, effect on ATP catabolism, 158
Ethylaminothiadiazole, effect on PP-ribose-P concentration, 149
Exercise:
  effect on muscle ATP, 122
  stimulation of purine nucleotide cycle, 128

F_0-adenosine triphosphatase, 30
F_1-adenosine triphosphatase:
  inhibitor protein, 31
  mechanism of ATP synthesis, 33-37
  subunits:
    roles of individual, 31-33
    structure, 31-33
Feedback inhibition, of purine biosynthesis *de novo*, 89
Fibroblasts:
  ATP catabolism, 105
  ATP synthesis, 105
Fluoride, effect on ATP catabolism, 99, 157
Fluorocortisone, effect on ATP metabolism, 121
Formycin:
  inhibition of PP-ribose-P synthetase, 155
  inhibition of purine nucleoside phosphorylase, 151
Free energy, standard, of ATP hydrolysis, 10-15
  charge effects, 14-15
  resonance stabilization, 14-15
Fructose, effect on ATP catabolism, 100
Fructose-1,6-bisphosphatase:
  deficiency disease, 141
  phosphofructokinase coupling, 62-63
Fructose-2,6-bisphosphatase, 63
Fructose-1,6-bisphosphate, regulation of glycolysis, 60-63
Fructose-2,6-bisphosphate:
  effect on fructose-1,6-bisphosphatase, 62-63
  effect on phosphofructokinase, 62-63
Fructose-6-phosphate, free energy of hydrolysis, 13

Glucagon, effect on ATP metabolism, 121
Gluconeogenesis, regulation of, 60
Glucose, effect on ATP catabolism, 99-100
Glucose-6-phosphatase, deficiency disease, 137, 141
Glucose-1-phosphate, free energy of hydrolysis, 12-14
Glucose-6-phosphate, free energy of hydrolysis, 12-14
Glucose-6-phosphate dehydrogenase, hormonal control, 121
Glutaminase, 11
Glutamine, regulation of ATP synthesis, 82-83
Glutamine synthetase, 11
  inhibitors, 154
Glyceraldehyde-3-phosphate dehydrogenase:
  half-sites behavior, 21-23
  reaction catalyzed, 21
  subunit structure, 21-23
Glycogen phosphorylase, effect of phosphorylation, 61
Glycogen storage disease, 137
Glycogen synthetase, effect of phosphorylation, 61
Glycolysis:
  inhibitors, 157
  regulation, 60-63
  substrate-level phosphorylation, 20-29
    anaerobic, 23

Gout:
and glucose-6-phosphate deficiency, 137
and hypoxanthine phosphoribosyl-transferase deficiency, 137-138
Guanazole, inhibition of ribonucleotide reduction, 146
Guanine nucleotides, synthesis, 86-87, 116
Guanosine triphosphate, regulation of ATP synthesis, 83
Guanylate, conversion to adenylate, 73, 75
Guanylate synthetase, 116

Hadacidin, inhibition of adenylosuccinate synthetase, 128, 155-156
Heart:
ATP catabolism, 102
ATP concentration, 5
factors affecting, 142
effect of adenosine, 123
Hemolytic anemia:
and adenosine deaminase hyperactivity, 141
and high ATP syndrome, 140
and inherited abnormalities of phosphorylation, 142
Hepatocytes, ATP turnover, 98
Hexokinase:
binding of substrates, 53-54
crystal structure, 52-54
deficiency disease, 142
kinetics, 49-50
mechanism of phosphoryl transfer, 50-51
phosphoenzyme, 50-52
reaction catalyzed, 49
High ATP syndrome, hereditary, 139-140
Histidine, biosynthesis, 117
Histones, effect of phosphorylation, 61
Hormones, effect on ATP metabolism, 121
Hydralazine, effect on ATP catabolism, 158
Hydrocortisone, effect on ATP metabolism, 121
β-Hydroxy-β-methylglutaryl-coenzyme A reductase, effect of phosphorylation, 61
Hydroxyurea, inhibition of ribonucleotide reduction, 146
Hypoxanthine:
catabolism, 96-97
drugs decreasing, 150
circulating, 120
conversion to adenylate, 73, 75
formation, 95
phosphorylation, 107
regulation of ATP catabolism, 107-108
reutilization for nucleotide biosynthesis, 129-130
Hypoxanthine phosphoribosyltransferase:
deficiency disease, 129-130, 136-138, 141
inhibitors, 156
mammalian cells lacking, 77
regulation of ATP catabolism, 107
tissue distribution, 105

Immunodeficiency disease:
and adenosine deaminase deficiency, 139
and purine nucleoside phosphorylase deficiency, 139
Indoramin:
effect on ATP catabolism, 150
Inosinate:
accumulation during exercise, 122
catabolism, 94-95
dephosphorylation, 86
circulating, 120
conversion to adenylate, 71, 86-87
conversion to guanine nucleotides, 87
conversion to xanthylate, 87
formation in ATP catabolism, 94-95
oxidation, 86
phosphorylation, 86
in purine nucleotide cycle, 107
regulation of ATP catabolism, 107
regulation of ATP synthesis, 86-87
synthesis:
de novo, 70-71, 74, 86-88
from guanylate, 73, 75
from hypoxanthine, 73, 75
Inosinate dehydrogenase, inhibitors, 147-148, 151
Inosinate nucleotidase, tissue distribution, 105
Inosine:
catabolism, 107
circulating, 120
regulation of ATP catabolism, 107
Inosine monophosphate cyclohydrolase, reaction catalyzed, 71
Insulin, effect on ATP metabolism, 121
Iodoacetamide, effect on ATP catabolism, 157
Idoacetate, effect on ATP catabolism, 99, 157
6-Iodo-9-tetrahydro-2-furylpurine, inhibition of hypoxanthine phosphoribosyl-transferase, 156

Isometamidium, effect on ATP catabolism, 158
Isoproterenol, effect on ATP metabolism, 121, 158

Kidney:
- ATP catabolism, 102
- ATP concentration, 5
- Na^+-K^+-ATPase, 42
- preservation for transplantation, 131-132

Kidney stones:
- and adenine phosphoribosyltransferase deficiency, 140
- and hypoxanthine phosphoribosyltransferase deficiency, 137
- and xanthine oxidase deficiency, 139

Lipase, adipocyte, effect of phosphorylation, 61
Liver:
- ATP catabolism, 102, 105
- ATP synthesis, 105

Lymphoblasts, ATP catabolism, 102

Magnesium, complex with ATP:
- coordination site, 15-18
- dissociation constants, 15-18
- effect on pK_a, 16-18

Manganese, complex with ATP:
- coordination site, 15-18
- dissociation constants, 15-18

6-Mercapto-9-(tetrahydro-2-furyl)purine, inhibition of hypoxanthine phosphoribosyltransferase, 156
Methionine sulfoximine, inhibition of glutamine synthetase, 154
Methotrexate, inhibition of dihydrofolate reductase, 154
Methylation reactions, 124-125
Methylene blue, stimulation of pentose phosphate pathway, 148
6-Methylmercaptopurine ribonucleoside, inhibition of amidophosphoribosyltransferase, 153-154
Methylthioadenosine phosphorylase, deficiency, 126
Microtubules, effect of phosphorylation, 61
Misonidazole, effect on ATP catabolism, 158
Mitochondria:
- ADP, ATP transport system, 59-60
- ATP concentration, 5
- ATP synthesis, 29-37
- F_0-ATPase, 30
- F_1-ATPase, 29-37
- role in maintenance of cytosolic ATP concentration, 59-60

Muscle:
- adenylate deaminase deficiency disease, 138-139
- ATP catabolism, 102
- ATP concentration, 5
  - maintenance of, 58-59
- contraction, 58-59
  - effect on ATP concentration, 122
  - sliding filament model, 48-49
- vasoregulation by adenosine, 123

Mycophenolic acid, inhibition of inosinate dehydrogenase, 147
Myosin:
- effect of phosphorylation, 61
- hydrolysis of ATP, 46-49
  - coupling to mechanical work, 48-49
  - rate-limiting step, 47-48
  - reaction sequence, 48
- structure, 46

Naftidrofuryl, effect on ATP catabolism, 150
Nebularin, inhibition of PP-ribose-P synthetase, 155
Nerves:
- ATP storage pools, 123-124
- effect of ATP, 122
- purigenic, 122

Neurohormones, release, 110
Neurotransmitters:
- ATP as, 122
- release, 110

Nicotinamide adenine dinucleotide, synthesis, 116
Nicotinamide adenine dinucleotide phosphate, synthesis, 116
Nicotinate adenine dinucleotide, synthesis, 116
p-Nitrobenzlthioinosine, and nucleoside loss from cells, 152
5′-Nucleotidase:
- inhibitors, 104, 108-109
- in intact cells, 108-109
- membrane-bound, 104
- reaction catalyzed, 95

Nucleus, ATP concentration, 5

Oligomycin, effect on ATP catabolism, 157
Ornithine decarboxylase, effect of phosphorylation, 61
Orotidylate decarboxylase, inhibitors, 149
Ouabain, inhibition of Na^+-K^+-ATPase, 42
Oxopurinol, inhibition of xanthine oxidase, 152

PALA, *see* N-(phosphonoacetyl)-L-aspartate
Pentose phosphate pathway, stimulation, 148
Phenazine methosulfate, stimulation of pentose phosphate pathway, 148
Phenyladenosine, inhibition of adenosine kinase, 156
Phenylalanine hydroxylase, effect of phosphorylation, 61
Phosphate, inorganic, effect on ATP catabolism, 99-100
Phosphate-pyruvate dikinase, site of ATP bond cleavage, 10
Phosphoadenosine phosphosulfate, synthesis, 117
Phosphoarginine, free energy of hydrolysis, 13
Phosphocreatine:
  free energy of hydrolysis, 13-14
  role in maintaining intracellular ATP concentrations, 58
Phosphoenol pyruvate, free energy of hydrolysis, 13-14
Phosphoenolpyruvate synthase, site of ATP bond cleavage, 10
Phosphofructokinase:
  deficiency disease, 142
  effect of phosphorylation, 60-63
  fructose-1,6- bisphosphatase coupling, 62-63
  regulation of, 60-63
Phosphoglycerate kinase:
  deficiency disease, 142
  mechanism of phosphoryl transfer, 24-25
  reaction catalyzed, 23
  structure, 23-24
Phosphohydrolases, dephosphorylation of ATP, 94-95
N-(Phosphonoacetyl)-L-aspartate, inhibition of aspartate transcarbamylase, 149
Phosphoribosyl aminoimidazolecarboxamide formyltransferase, reaction catalyzed, 71
Phosphoribosyl aminoimidazole carboxylase, reaction catalyzed, 70
Phosphoribosyl aminoimidazole succinocarboxyamide synthetase, reaction catalyzed, 71
Phosphoribosyl aminoimidazole synthetase, reaction catalyzed, 70
Phosphoribosyl formylglycinamide synthetase, reaction catalyzed, 70
Phosphoribosyl formylglycinamidine synthetase, inhibitors, 153
Phosphoribosyl glycinamide formyltransferase, reaction catalyzed, 70
Phosphoribosyl glycinamide synthetase, reaction catalyzed, 70
Phosphoryl transferases, site of ATP bond cleavage, 10
Phosphorylation potential of cells, 98
Platelets:
  aggregation, 110
  ATP storage, 123
Polyadenosine diphosphate ribose:
  structure, 117
  synthesis, 117
Polyamines:
  synthesis, 125-126
Potassium, transport, 41-42, 45. *see also* Adenosine triphosphatase (Na^+ -K^+)
PP-ribose-P:
  drugs affecting, 149-150
  regulation of ATP synthesis, 75, 81-82, 88
  synthesis, 81-82
  utilization, 149
PP-ribose-P synthetase, 81
  abnormalities, 136
  hyperactivity, 136
  inhibitors, 78-79, 108, 155
    end product, 89
  kinetics, 78-79
  regulation by ATP, 89-90
  regulation of ATP synthesis, 78-79
  subunit structure, 79
Protein kinases, and metabolic control, 60
Proteins, adenylation, 118
Proton pump, in mitochondrial membrane, 30-31
Psicofuranine, inhibition of PP-ribose-P synthetase, 155
Purigenic nerves, 122
Purine nucleoside phosphorylase, 105
  deficiency disease, 139
  inhibitors, 151
  reaction catalyzed, 95-96

regulation of ATP catabolism, 107
tissue distribution, 105
Purine nucleotide cycle, 126-129
Purines:
biosynthesis following anoxia, 130-131
catabolism, 96-97
circulating, effect on ATP metabolism, 120-121
dietary, effect on ATP metabolism, 118-120
Pyrazofurin, inhibition of orotidylate decarboxylase, 149
Pyrophosphoryl transferases, site of ATP bond cleavage, 10
Pyrroline-5-carboxylic acid, stimulation of pentose phosphate pathway, 148
Pyruvate dehydrogenase, effect of phosphorylation, 61
Pyruvate kinase:
deficiency disease, 142
effect of phosphorylation, 61
mechanism of phosphoryl transfer, 25-28
nucleotide binding site, 26-28
reaction catalyzed, 25-26

1-β-D-Ribofuranosyl-1,2,4-triazole-3-caboxamide, inhibition of purine nucleoside phosphorylase, 151
Ribonucleases, in RNA turnover, 129
Ribonucleic acid:
dietary, effect on ATP metabolism, 119-120
synthesis, 115
inhibitors, 146
turnover, 129
Ribonucleotide reductase:
inhibitors, 146-147
reaction catalyzed, 115
Ribose-5-phosphate:
synthesis, 81-82
utilization, 81-82
Rotenone, effect on ATP catabolism, 99, 157

S-adenosylhomocysteine, in methylation reactions, 124-125
S-adenosylmethionine:
effect on ATP concentration, 157
intracellular concentrations, 124-125
in methylation reactions, 124-125
in polyamine synthesis, 125-126
synthesis, 116
Sangivamycin, inhibition of RNA synthesis, 146
Sliding filament model, of muscle contraction, 48-49
Sodium, transport, 41-42, 45. *see also* Adenosine triphosphatase (Na^+ -K^+)
Spermidine, synthesis, 125-126
Spermine, synthesis, 125-126
Substrate-level phosphorylation, 20-29
Substrate synergism, 50-51
Succinyl-coenzyme A synthetase:
nucleotide specificity, 28
phosphoenzyme, 50
reaction catalyzed, 28-29
subunit structure, 29
Sympathetic nerve vesicles, ATP storage, 123

Tetradydrofolate coenzymes, synthesis, inhibition, 154-155
Thiamine, synthesis, 86
Thiosemicarbizones, inhibition of ribonucleotide reduction, 146
Thyroid-stimulating hormone, effect on ATP metabolism, 121
Tissue, preservation, 131-132
Tolbutamide, effect on ATP metabolism, 121
Tricarboxylate cycle, substrate-level phosphorylation, 20-29
Triosephosphate isomerase, deficiency disease, 142
Tropomyosin, effect on phosphorylation, 61
Troponin I, effect of phosphorylation, 61

Urate oxidase, reaction catalyzed, 96-97

Virazole, inhibition of inosinate dehydrogenase, 147

Xanthine, catabolism, 96-97
Xanthine dehydrogenase, reaction catalyzed, 96-97
Xanthine oxidase:
deficiency disease, 139
in hypoxanthine metabolism, 108
inhibitors, 152
reaction catalyzed, 96
tissue distribution, 108, 120
Xanthylate aminase, 116
Xanthylate nucleotides, synthesis from inosinate, 86-87
Xylosyladenine:
inhibition of PP-ribose-P synthetase, 155
inhibition of RNA synthesis, 146